The Modern Crisis

The Modern Crisis

Murray Bookchin

Foreword by Andy Price

Also by Murray Bookchin

Our Synthetic Environment

Crisis in Our Cities

Post-Scarcity Anarchism

Limits of the City

The Spanish Anarchists

Toward an Ecological Society

The Ecology of Freedom

Remaking Society

The Philosophy of Social Ecology

From Urbanization to Cities

Defending the Earth

Which Way for the Ecology Movement?

To Remember Spain

Re-enchanting Humanity

Anarchism, Marxism, and the Future of the Left

Social Anarchism or Lifestyle Anarchism

The Third Revolution (Volumes 1–4)

Social Ecology and Communalism

The Next Revolution

The Modern Crisis (Third Edition)
Edited by Debbie Bookchin

Foreword © 2022 Andy Price
All rights reserved.
Third Edition © 2022 The Bookchin Trust

ISBN 978-1-84935-446-2
E-ISBN: 978-1-84935-447-9
LCCN: 2021944522

AK Press   AK Press
370 Ryan Avenue #100   33 Tower Street
Chico, CA 95973   Edinburgh, EH6, 7BN
USA   Scotland
www.akpress.org   www.akuk.com
akpress@akpress.org   akuk@akpress.org

Please contact us to request the latest AK Press distribution catalog, which features
books, pamphlets, zines, and stylish apparel published and/or distributed by AK
Press. Alternatively, visit our websites for the complete catalog, latest news, and
secure ordering.

Cover design by John Yates, www.stealworks.com
Printed in the United States of America on acid-free paper

About this edition:
The original edition of *The Modern Crisis* was published in 1986 and included four
essays: "Rethinking Ethics, Nature, and Society," written specifically as an introduction to the book in 1985; "What is Social Ecology?" delivered as a lecture-seminar at
the Goethe University Frankfurt in 1984. It was revised for publication in a textbook
in 1993, and lightly revised in 2005, and again in 2022 for this edition; "Market
Economy or Moral Economy," based on a keynote address Murray Bookchin gave
at the annual convocation of the New England Organic Farmers' Association in
June 1983; and "An Appeal for Social and Ecological Sanity," which was written as
a pamphlet for the Institute for Social Ecology in 1983. For a second edition of *The
Modern Crisis*, published in 1987, Bookchin added the essay, "Workers and the Peace
Movement," which was originally written in July 1983 and appeared in Bookchin's
occasional brochure *Comment*. This new edition contains a newly shortened version
of "An Appeal for Social and Ecological Sanity." It adds a sixth essay, "Radical
Politics in an Era of Advanced Capitalism," first published in Bookchin's *Green
Perspectives* in 1989.

*For my son Joe
and for
Bob and Marilyn Bookchin*

CONTENTS

Foreword
Andy Price

It was back in 1999 and my university friends and I were cruising around the new-fangled computer network with which we were barely familiar—known at the time as the World Wide Web. News had reached us of a new protest movement, taking place outside institutions equally unfamiliar to us: The World Bank, the International Monetary Fund. The groups protesting were even more mysterious still: The Direct Action Network, Jubilee 2000—and what, exactly, was a Zapatista? Moreover, what were they doing? Protesting? Against what?

As with most people who had come of age in the West of the 1990s, political protest had become distant to us. Whether or not one was a student of politics, the economic and political triumphalism of the 1990s was difficult to avoid: the Cold War was over; the West had won. Capitalism was the only show in town now. It was not just an economic system; it was the new foundational ideology of the day. Even the old opponents of unfettered capitalism and its buccaneering free market in the West had embraced this new economic reality—Clinton in the U.S., Blair in the U.K., Schroeder in Germany. In this setting, the shouts of protest coming from Seattle in 1990 couldn't have been starker: weren't we all now living the high life in the "end of history"?

With this question in mind, I hit the university library, and looked for anything that could help explain these protests. There, I found an obscure-looking text, complete with a graffiti-style font that made it look like either a book on art or a book on breakdancing. That book was *The Modern Crisis,* by Murray Bookchin, a collection of essays on capitalism, society, the economy, and the environment, written fifteen years before Seattle; and yet a collection that pointed clearly to where the groups and their ideas at Seattle could have emerged.

"The fact is that our present market economy is grossly *immoral,*" Bookchin declares in his far-sighted essay "Market Economy or Moral Economy." "The economists have *literally* 'demoralized' us and turned us into moral cretins," he continues (p. 62 below). As the activists in Seattle had come to realize, were we not all living in the triumphant, golden age of capitalist economics after all. Rather, the cutthroat competition of unregulated market capitalism was now becoming our social mores *as such*; "our social map has been completely taken over by the market," Bookchin tells us (p. 63 below).

Bookchin offered here a nascent but excoriating critique of the effects this transformation was having: "we not only buy and sell our labor power" in this new *market society*, we also "buy and sell our own neuroses, anomie, loneliness, spiritual emptiness ... we buy and sell the outward trappings of personality" (p. 65 below). He also pointed toward a different way of conceiving of economic activity. Preempting David Graeber's work in *Debt* by two decades, Bookchin reminded us that historically, economic activity had always, in some sense, been imbued with notions of moral commitment: the terms "good" and "goods" point to the early conceptions of what it is to produce and exchange in society; early philosophy and economics point to the ties that came from each individual's economic activity. Whilst there is no singular, historical model for what makes a moral economy, the historical record is replete with ideas and individual practices that point toward a morality in economics (one that Graeber even saw as the bedrock of credit and debt), which in turn points to the tantalizing possibility of re-moralizing economic activity in the present day.

Crucial also here—and so far-reaching when it was first penned in 1984—is Bookchin's critique of the handmaiden of this transformation of a society to an economic system: that unalloyed bastion of "progress," the modern nation-state. Bookchin pulls no punches in debunking the myth of statehood, and it is worth quoting from his opening essay "Rethinking Ethics, Nature, and Society" at length here to see just how radical this essay was:

> The nation-state makes us less than human. It towers over us, cajoles us, disempowers us, bilks us of our substance, humiliates us—and often kills us in its imperialist adventures. To be a citizen of a nation-state is an abstraction that removes us from our lived state to a realm of myth, clothed in the superstition of a "uniqueness" that sets us apart, as a national entity, from the rest of humanity—indeed, from our very species. In reality, we are the nation-state's victims, not its constituents—not only physically and psychologically, but also ideologically. (p. 30 below)

But if *The Modern Crisis* landed an early blow against a dawning economistic age, it wasn't the impact of this transformation on individuals that was Bookchin's only concern. Across the other articles collected here, he outlines the growing prospect of ecological collapse. Of course, by the mid-1980s, so called "green issues" had become almost mainstream discussion points: pollution, the danger of nuclear power, habitat loss. But here, Bookchin went much further: "Economic, ethnic, cultural, and gender conflicts, among many others, lie at the core of the most serious ecological dislocations we face today—apart, to be sure, from those that are produced by natural catastrophes," he argues in his seminal essay "What is Social Ecology?" Debunking again the triumphalism of the closing decades of the twentieth century, Bookchin continues:

> Unless we realize that the present market society, structured around the brutally competitive imperative of "grow or die," is a thoroughly impersonal, self-operating mechanism, we will falsely tend to blame other phenomena—such as technology or

population growth—for growing environmental dislocations. We
will ignore their *root* causes, such as trade for profit, industrial
expansion for its own sake, and the identification of progress
with corporate self-interest. In short, we will tend to focus on the
symptoms of a grim social pathology rather than on the pathology
itself, and our efforts will be directed toward limited goals whose
attainment is more cosmetic than curative. (p. 36 below)

However, perhaps most importantly in *The Modern Crisis*, from a
theoretical point of view at least, is the way that, even in the midst of
all of this human-induced destruction, Bookchin still places humanity at the center—not just as the *cause célèbre* of this destruction, but
also as its solution, if humanity itself can only be recovered first.

In this sense, Bookchin brings together the two themes, not only
of this collection, but of his life's work: that ecological problems are
first and foremost social problems; thus their solutions can only be
social too. And the further we reduce society to a set of economic
rules, the worse these problems and the search for their solutions
become. Instead, Bookchin says, we must fully divest society of the
hierarchical structures that underpin capitalism and the nation
state.

[Social Ecology] challenges the entire system of domination
itself—its economy, its misuse of technics, its administrative apparatus, its degradations of political life, its destruction of the
city as a center of cultural development, indeed the entire panoply of its moral hypocrisies and defiling of the human spirit—
and seeks to eliminate the hierarchical and class edifices that
have imposed themselves on humanity and defined the relationship between nonhuman and human nature. (pp. 54–55 below)

These observations elaborate a theme that Bookchin has honed
throughout his work: that the ecological devastation humans have
caused is perhaps at its worst when it comes to the devastation we
wreak on ourselves. This is not because humanity is inherently more
important than other species; rather, it is based on the irreplaceable

loss we will suffer if the only species capable of reversing that destruction destroys its opportunity to do so.

It is for this reason that I remain as inspired by *The Modern Crisis* as when I first picked it up: across this collection was a clarion call for just how much trouble we were in, socially and ecologically, published at the dawn of a quarter century of American and Western triumphalism. Yet, while offering a devastating critique of humanity's crisis, Bookchin simultaneously insists on and celebrates the fact that humanity has the tools and capability to rescue itself from that crisis. In that sense, it is a characteristically Bookchin work: an unflinching critique of the present, but suffused with hope that humanity can change—a theme that is as crisp and vivid here as found anywhere in his work.

March 2022

Rethinking Ethics, Nature, and Society

This introductory essay and the five additional articles that make up *The Modern Crisis* are guided by the view that our ideas and our practice must be imbued with a deep sense of ethical commitment. We must recover an image of the public good in a world that increasingly makes its choices between one "lesser evil" and another. Such an endeavor can easily slip into self-righteous sermonizing if we do not examine certain basic assumptions that distinctly mark the present era.

Most of the "isms" we have inherited from the past—liberalism, socialism, syndicalism, communism, capitalism—are rooted in the crude notion that human beings act almost exclusively from self-interest. This notion unites such widely disparate and immensely influential thinkers as Adam Smith, Karl Marx, and Sigmund Freud, not to speak of a huge bouquet of liberals, socialists, and self-styled "libertarians" (more properly proprietarians—the acolytes of Ayn Rand) in a common vision of human motivation and social behavior. Some time after the Enlightenment and the Victorian era that followed it, ethical approaches to freedom, self-consciousness, and harmony began to give way to appeals for a "scientific," presumably "materialist," approach to a social reality grounded in egotism and the picture of a self-serving, indeed, avaricious human nature.

Even so libertarian a visionary as Mikhail Bakunin, the fiery voice of nineteenth-century anarchism, echoes Marx and many radicals of his day when he militantly declares that, "Wealth has always been and still is the indispensable condition for the realization of everything human."[1] Let there be no mistake that these words are directed simply against any particular body of ideology or dogma. They reflect a widely accepted tendency in ideas that steadily reduced earlier pleas for freedom to "ideological" expressions of (read: apologias for) free trade; pleas for consciousness to "mystified" expressions of class interest; and ideals that called for a harmonious world to attempt to conceal a "historical" need for social discord—whether in the form of swashbuckling competition in the economic sphere or armed conflict in the social sphere.

What deeply disturbs me is the extent to which this image is so integrally part of our prevailing market economy—the profit-oriented capitalistic world men like Bakunin and Marx so earnestly opposed. The "embourgeoisment" of the working class, which troubled Bakunin in the 1870s, has expanded to embrace virtually all the radical "isms" we have inherited from the last century. Our bourgeois society has carried its own ideological baggage, filled with notions of self-interest, into the mental world of its opponents.

As a result, traditional radicalism in nearly all its forms has become the alter ego of traditional capitalism. In a social universe where rival ideologies—be they conservative, liberal, or radical—root seemingly conflicting interpretations of social development in egotism, we are faced by a good deal more than the problem of ideological and psychological cooptation. The power of self-interest, whether we choose to call it "class interest" or private interest, becomes so much a part of the received wisdom of our period that it unconsciously shapes all our ideological premises. Causes like "socialism," "liberalism," or even "anarchism" become wedded to a distinctly bourgeois outlook in its crudest and most elemental sense even as they try to advance ideals that serve the public welfare.

The practical results of these assumptions are insidiously corrosive. Not only is every moral challenge to self-interest indulgently dismissed as "naïve," "idealistic," and "utopian," but many people

who are principled opponents of the prevailing order easily slip
into "strategies" that appeal to specific interests in their crassest
forms. "Bread-and-butter" issues, important as they surely are, are
given almost embarrassing priority over moral and even long-range
material ones. Thus, the runaway growth that threatens to undermine
the environment, the centralization of power that inches society
toward an authoritarian State, and increases in armaments production that create jobs at the expense of the community's autonomy
from the military industrial complex are sanctified by the immediate
needs of workers for employment.

A cozy adaptiveness to the seeming imperatives of "realism"
is widely prevalent among people who avow a moral adherence to
freedom. Radicalism becomes radically schizophrenic: a fragile,
unworldly life of ideals coexists with a stolid, very worldly practice spawned by opportunism. The hidden agenda, with its visible
"superstructure" composed of pitches to self-interest and concealed
"base" of lofty principles reserved for private discussion—a "just-
between-us" kind of socialism and liberalism that presumably balances out a public life of manipulative calculation—becomes the
rule of our time. Given this chasm between principle and practice,
we can justly ask what has happened to the "self" that structures its
ideology and activities around self-interest on the one hand and an
emancipatory self-consciousness on the other. The modern crisis, in
effect, includes a serious crisis in radicalism itself and in many ideological movements that try to resolve the dislocations of our time.

Any project that seeks to deal with this crisis must also be a
project that rehabilitates the prevailing image of human motivation. Indeed, it is an endeavor that must go still further: it must try
to separate radicalism (or, if you choose, the liberatory movement
in all its forms) from its deep roots in our market society and the
vicious mentality it breeds. Whether in socialism, anarchism, or liberalism, moral issues must be raised again to remove a toxic legacy
of realpolitik that has shunted them aside from their centrality in
struggles for human emancipation. I know of nothing more warped
than attempts to come to terms with a thoroughly irrational "reality" and a "realism" that increasingly asks us to choose between

the alternatives of nuclear immolation and ecological disaster. This dilemma, which is the height of *unreality* in Hegel's famous equation of the "real" with the "rational," forms the "big picture" of our time, or what sociologists pompously call its "macro-problems." Entering deeply into the factors that have produced the dilemma is a moral issue: the conviction that every benefit must be "purchased" by a risk—in short, that, for every advancement, humanity must pay a penalty. Why this should be so remains inexplicable unless good is indissolubly wedded to evil, in which case ethics conceived as a pursuit of virtue becomes a contradiction in terms.

What this "benefit-versus-risk" formula has produced, however, is a monumental apologia for all the ills of our time, a justification for the existence of evil as such and its entry into the core of life in the modern world. When good and bad are no longer placed in opposition to each other but rather adjusted to *coexist* with each other as though they were integrally part of the same phenomenon, ethical standards dissolve into techniques for accommodation to things as they are.

Even a moral movement to redeem movements for social change becomes a mere problem of calculation—specifically, of the risks we must incur in matters of principle in exchange for the benefits we wish to gain in matters of practice. This is a typical bourgeois calculation, which every entrepreneur faces when he or she enters the marketplace jungle, a calculation that has become an issue in ethics precisely because moral behavior in the marketplace is all but impossible. That a benefit-versus-risk mentality has become the common coin of everyday discourse attests to the all-pervasiveness of a market economy. It exists no less in modern radicalism than in modern commerce. In both cases, the sweet smell of success, however opportunistically achieved, tends to win over the odium of possible failure in matters of principle.

There is a cruel, indeed, ironic justice to this calculation. Each risk that a benefit incurs, each "lesser evil" bought at the expense of principle, ultimately yields a universe of risks and evils that by far surpasses the original pair of choices that lead us to this ill-conceived strategy. In politics, Weimar Germany between World War I and II provides us with the classic example of this ethical devolution. Faced with paired choices of "lesser evils" or "benefits-versus-risks" from one election to another, German Social Democracy played a gamble with destiny that led it from the prospect of revolution, which it unscrupulously aborted, to choices between a moderate Left and a tolerant Center, a tolerant Center and an authoritarian Right, and finally an authoritarian Right and totalitarian Fascism. The results, of course, are history, a history presumably tucked away in the past. But the strategy is painfully contemporary. In daily life, this devolution is an ongoing process that has become so automatic that it no longer appears to be a matter of choice. So complete is the surrender of ethics to the mere process of functioning, of principle to the mere routine of survival, that it has become an unthinking series of operations at the most molecular level of life.

One can easily blame people for becoming caricatures of their lost humanity. Everyday life has steadily acquired almost bovine characteristics. Society is little more than a pasture and people a herd grazing on a diet of trivialities and petty pursuits. The price we pay for this repellent reduction of humans to domesticated, shepherded, and unthinking beings is costly beyond imagination. The insults our elites inflict on their subordinates by cajoling them do not solve the problems of our times, but simply remove them from public purview.

Hence, the reinstatement of an ethical stance becomes central to the recovery of a meaningful society and a sense of selfhood, a realism that is in closer touch with reality than the opportunism, lesser-evil strategies, and benefits-versus-risk calculations claimed by the practical wisdom of our time. Action from principle can no longer be separated from a mature, serious, and concerted attempt to resolve our social and private problems. The highest realism can be attained only by looking beyond the given state of affairs to a vision of what *should* be, not only what *is*. The crisis we face in human subjectivity as well as human affairs is so great, and its received wisdom is so anemic, that we literally will not *be* if we do not realize our potentialities to be more than we are.

The scope of the modern crisis is reconnoitered in the pieces that make up this book. My intention here is not to review the problems that have formed it. That would be redundant. I would like, how-ever, to examine their depths and their connections to each other, to go beneath the surface of the book and explore certain cohering thoughts from which we can develop a meaningful whole.

If we desperately need an ethics that will join the ideal with the real and give words like "realism" a richer, more rational meaning than they have, then we are faced with a traditional dilemma. How can we objectively validate ethical claims in an era of moral relativ-ism when good and bad, right and wrong, virtue and evil, even the selection of strategies for social change are completely subjectivized into matters of taste or opinion? The overstated claim that what is good for a highly personalized "me" may not be good for an equally personalized "you," speaks to the growing amorality of our time. Accordingly, such a moral relativism (I can hardly call it a relativ-istic "ethics" without debasing the very meaning of the word *ethics*) has acquired the sanctity of a constitutional precept in our system of government. It has become the standard by which to determine the criminality of behavior and the guiding principles of diplomacy, reli-gion, politics, and education, not to mention business and personal affairs. The subjectivization of behavioral precepts reflects the uni-versal opportunism of the time; its emphasis is on operational ways of life as distinguished from philosophical ones, especially on ways to survive and function, rather than on ideals imbued with meaning.

That moral relativism can deliver us to a totally noncritical view of a world in which mere taste and fleeting opinion justify any-thing, including nuclear immolation, has been stressed enough not to require further elucidation. If mere opinion suffices to validate social behavior, then the social order itself can be validated sim-ply by public opinion polls. Hence, whether capital punishment is "right" or "wrong" ceases to be an ethical question about the sanctity of life. The issue becomes a problem of juggling percentages, which may justify the slaughter of homicidal felons during one year and their right to live during another. Whether the figures of our polls go up or down can decide whether a given number of people will be put to death or not. Carried to its logical conclusion, this personalistic, operational view of morality can justify a totalitarian society, which abolishes the very claims of the individual. It was not from a sense of irony or perversity that visitors to Mussolini's Italy in the 1920s applauded a fascist regime because Italian trains operated on time. The efficiency of a social system and mere matters of personal convenience were identified with its claims to be the embodiment of the public welfare.

To exorcise moral relativism, with its distasteful extensions into a politics of lesser-evils and a practice structured around risk-versus-benefit calculations, is a vexing problem indeed. The converse of a radical moral relativism is a radical moral absolutism that can be as totalitarian in its power to control as its relativistic opposite is democratic in its power to relax. Both live in a curious intellectual symbiosis: the seeming pluralism of a moral democracy has been known to encompass a fascistic ethics as easily as an anarchic one—which raises the question of how to keep a democracy from voting itself out of existence.

Suffice it to say that moral absolutism is neither better nor worse than the *concrete* message it has to offer. An ethics grounded in ecology can yield a salad of "natural laws" that are as tyrannical in their conclusions as the chaos of a moral relativism is precariously wayward. To appeal from ecology to God is to leap from nature to supernature, that is, ironically, from the human subject as it exists in the real world to the way it exists in the imagination. Religious precepts are the products of priests and visionaries, not of an objective world from which we can gain a sense of ethical direction that is neither the commanding dicta of "natural law" on the one hand nor supernatural "law" on the other. We have learned only too well that Hitler's "blood and soil" naturism, like Stalin's cosmological "dialectics," can be used as viciously as notions of "natural law" (with all their Darwinian connotations of "fitness to survive" and "natural selection") to collect millions of people in concentration camps, where they are worked to death, incinerated, or both.

Indeed, the suspicion surrounding the choice of nature as a *ground* for ethics is justified by a history of nature philosophies that gave validity to oligarchy (Plato), slavery (Aristotle), hierarchy (Aquinas), necessity (Spinoza), and domination (Marx), to single out

to become a fully mature and creative adult. This notion, in any case, is a message of freedom, not of necessity; it speaks to an immanent striving for realization, not to a predetermined certainty of completion. What is potential in an acorn that yields an oak tree or in a human embryo that yields a mature, creative adult is equivalent to what is potential in nature that yields society and what is potential in society that yields freedom, selfhood, and consciousness.

I find it odd that social ecology—the most organic of our social disciplines—is often discussed in strictly reductionist and analytical terms. Indeed, there is a strong tendency to *collect* ideas rather than *derive* them, to disassemble or reassemble them as though we were dealing with an automobile engine, rather than explore them as aspects of a process. If recent cosmological theories about the universe are sound, the notion that it originated from a pulse of energy does not mean that all "matter" can henceforth be reduced to energy. Rather, the ecological thrust of that originating pulse has been to elaborate and differentiate itself, forming subatomic, atomic, molecular, and finally, richly elaborated and ever more complex inorganic and organic forms. Moreover, without introducing any notion of predetermination and teleology into our ways of thinking, each form can best be understood as *emerging* out of its predecessor—the later and more complex generally incorporating the earlier and simpler ones, whether internally or as part of a community.

This biological or organic way of thinking—which in no way conflicts with the proper use of mechanical or analytical forms of thought but rather encompasses them—is strangely lacking in many socially oriented schools of ecology. I still encounter schools that tend simply to inventory energy on one side and "matter" on an opposing side instead of deriving the latter from the former. Similarly, biocentric values are opposed to anthropocentric, the objective world of things is opposed to the subjective world of ideas, the strictly natural is opposed to the strictly social. We would do well to ask if they are in conflict with, or reducible to, each other—indeed, if we are talking about them in a thoroughly rounded manner when we render them so one-sidedly and simplistically. Even the "horrid" words "anthropocentricity" and "humanism," so disdained by many socially oriented ecologists, raise the question of whether humans have their own special place in nature with all their uniqueness and their own distinctive contribution to the whole.

We are very much in need of organic, more precisely, really dialectical ways of process-thinking that seek out the potentiality of a later form in an earlier one, that seek out the "forces" that impel the latter to give rise to the former, and that absorb the notion of process into truly evolutionary ways of thought about the world. Until this organic mode of thought is brought to ecophilosophy and applied in a sensitive, richly nuanced, and rounded manner, our attempts to reflect deeply on ecological problems will tend to be painfully superficial and incomplete.[4]

I can give no "definition" of social ecology that excludes the totality of these concepts, unified by the process that produces them. Just as Herbert Marcuse "defined" capitalism as all that appears and is worked out in the three volumes of Marx's *Capital* (and, I would be inclined to say, a good deal more), so social ecology is what I have tried to detail in the passages above and a good deal more. At a time when social ecology, once a rarely used term which I chose to express the ideas contained in this book, has become increasing familiar, it would be well to bear its specific meaning in mind. To divest this term of its liberatory, processual, and ethical content is to completely hybridize utterly antithetical concepts. Our age, which faces intellectual suffocation because of a massive denaturing of language into buzz words and a degradation of concepts into simplistic definitions, is guided too often by academic fashions and media-created fads. To resist this melding of words and concepts into a goulash of ideas is to resist the degradation of mind itself. Given the unabated production of fashions, fads, and superficial ideas, we have compelling reasons to fear for our spiritual degradation as well as our ecological degradation.

The very word *social* that is added to the word *ecology*—in contrast to the more commonly used term, "human ecology"—is meant to emphasize that we can no more separate society from nature than we can separate mind from body. If nature provides the ground for an ethics that has an objective ancestry in evolution's thrust toward freedom, selfhood, and reason, so too nature provides the ground for the emergence of society.

Here again we must exhibit the utmost delicacy in the treatment of ideas. Animal communities are not societies. Whatever else they have, they do not form those uniquely human contrivances we call institutions, which systematically and often purposively organize relationships among people along kinship lines in tribal societies, political ones in cities, and statist ones in nations and empires. Just as we must presuppose the emergence of municipal institutions that, in classical Athens, produced the *polis* or so-called city-state, a community that gave due recognition to an extraordinarily intelligent figure like Pericles, so we must presuppose the emergence of national institutions that gave virtually absolute power to idiots like Louis XVI of France and Nicholas II of Russia. In either case, the leadership and power conferred on these individuals had little to do with their physical strength (the usual source of so-called dominance in animal groups) and, with the exception of Pericles, their mental capacities. Rather, their authority came from well-organized, carefully structured, and historically developed agencies: armies, bureaucracies, police, or, in Athens's case, a large "town meeting" called the *ecclesia*. We encounter nothing like these agencies in the animal world. The so-called castes we find among "social" insects like bees and termites are genetic in origin, not contrived. By their rigidity they are fixed in ways that stand sharply at odds with human forms of organization, which are repeatedly altered by reforms or simply overthrown by revolutions.

The ways in which human societies have an ancestry in animal communities are too complex to examine here. One thing should be clarified, however: the two are not the same. Indeed, with the elaboration of human societies beyond the most elementary forms of scavenging and food gathering, we can see the emergence in humanity of two societies: the sororal society of women and the fraternal society of men. Perhaps the most underlying factor that makes for this duality in tribal communities is a division of *functions*—not simply a division of "labor." For it is more than work that separates the sexes into two fairly delineable social groups: it is culture itself. Women and men develop their own distinctive lifeways, modes of expression, behavioral traits, values, rituals, sensibilities, even deities, myths, and traditions, as well as styles of work, not only forms of work. Home, garden, cleaning, food preparation, parenting, and many other functions constitute a complete domain that is distinctively a woman's realm. Man's domain consists of hunting, "politics" (to misuse a word that requires explanation), and, where it exists, the men's house, into which all the males withdraw after puberty to form a separate community of their own.

The two societies have existed, whether in well-developed or virtually vestigial form, from time immemorial. Initially, they were in a reasonable balance with each other, a balance that was notable for the absence of domination by either sex. Only later do we begin to encounter a relationship marked by male dominance. Hence hierarchy, where it existed at an early point in social development, was the exception rather than the rule. Difference, I must emphasize, must not be mistaken for dominance or submission, a problem that continually recurs in a world that organizes all differentia into an order of "one-to-ten" in everyday life and institutionalizes them into a "chain of command" in social and economic areas of life.

We also have a difficult time recognizing the very existence of these two societies and are even more troubled in our search for the precise reason "why" men came to dominate women. Accordingly, we look for gimmicks that explain this shift from equality between the sexes—an equality whose existence is often denied—to a condition clearly marked by male domination or the prevalence of a "man's world." The mechanical, often reductionist, nature of these explanations is found most clearly in the view advanced by Dorothy Dinnerstein, notably, that women who provide succor for children are also the earliest source of denial to them. Some of Dinnerstein's admirers, carrying this view to its extremes, argue that the feelings of denial that stem from woman's "monopoly of parenting" turn her into an object of hatred as well as love, and that this can be relieved by shared parenting by men and women. I find this explanation of misogyny, patriarchy, and even hierarchy highly questionable. It is not only too simplistic and reductionist, but also fails to account for the many variations in women's status that appear in societies where women continue to exercise a "monopoly of parenting" and, by any standards, are often far more loved by the young than are their fathers.

We come, here, to a highly revealing prejudice in the way our society today looks at the past, a difference that highlights the distinction between ecological and conventional ways of assessing human development. Underlying the widespread notion that woman has always lived in a "man's world" is a distinct bias that raises the male realm of civil society over woman's domestic world. Nourished by the enormous influence politics and statecraft have had on modern thinking, we tend to assume that civil society is always more important than domestic society, that "affairs of state" have primacy over the affairs of the household.

In tribal societies, this prejudgment of what is important and what is not is often pure fiction. Woman, whose food gathering and gardening activities often provide as much as eighty percent of the biomass consumed by hunter-gatherer societies, enjoys an economic eminence that certainly equals or even exceeds that of men. If we bear in mind that band and tribal communities are mainly domestic societies, the enormous importance we assign to civil society is largely a modern prejudice. Indeed, one could advance very persuasive arguments to support the claim that these societies were originally a woman's world and that women enjoyed a status that was "superior" to that of men.

Actually, we have no reason to suppose that the preeminence of a "woman's world" over a man's was marked by the dominance of females over males. Notions of dominance and submission have a very checkered history—even when applied to animals—that cannot be unraveled here. It is my own conviction that the *expansion* of the male's civil sphere, not its *a priori* supremacy at the very beginnings of the development of human social forms, explains the increasing supremacy of men over women—a view, I may add, that does not exclude the operation of many complex psychological, possibly even biological, factors. These may have reinforced patriarchy, but they do not in themselves explain "why" it emerged. It seems more likely, in my view, that civil society began to encroach upon domestic society and increasingly produced a "man's world" because of population pressure, the evolution of warfare (itself a very complex process), technological changes, and the slow reworking of early egalitarian societies by the elders of the community, its shamans, and, perhaps most decisively, its warriors.

The fallacies that the problem of male domination has generated, however, are significant indicators of the biases that interfere with an open study of social development. Difference, it cannot be emphasized often enough, *does not by itself yield hierarchy* or even a certain measure of dominance. The existence of two societies—male and female—does not justify a conclusion that one exercised supremacy over another. Finally, the fact that the male's civil society did achieve supremacy over the female's does not mean that it had to do so. The ability of social ecology to distinguish differentiation from domination, indeed, to visualize variation as part of wholeness rather than pyramidally, raises an important alternative. It opens the way to a sensibility that emphasizes harmony over antagonism and fosters a life-affirming ethics, objectively grounded in a fecund nature, which places a premium on variety, uniqueness, and the ability of life forms to complement each other in forming richer and ever-more creative wholes.

That hierarchy and domination did develop is another thesis, largely implied, that runs through the articles in this book. But here, too, we encounter problems that reveal an appalling wrongheadedness: the tendency, generated by so many liberals and Marxists, to anchor all relations in self-interest and economic motivation. I refer to the reduction of hierarchy to classes and of domination to exploitation. Hierarchy has a much broader meaning than class, a strictly economic relationship that Marx, quite properly, rooted in the ownership of property, the control of technology, and the various ways of operating the means of production. What is important, indeed crucial, about this distinction is that hierarchy preceded the emergence of classes and may long survive them if we are not mindful of its far-reaching implications. Hierarchy goes beyond the workplace into the very cradle, so to speak, indeed, into the socializing process, where infants are taught to deal with "otherness" as potentially hostile, as "objects" to be controlled. It is the breeding ground of those primary distinctions based on gender that are inculcated in the young. Finally, it teaches the young to accept their place in a social pyramid that reaches from the family to the summits of their adult lives. Not only are families, peer groups, educational institutions, religious centers, and the community, at large, schools for hierarchy as they are now constituted, so, too, are the workplaces and, in no small part, the technologies that place people in their service. A classless society may emerge that poses no challenge to any of these social forms. A nonhierarchical society, challenging the most archetypal components of social life, indeed, the socializing process itself, goes much further. It poses the need to alter every thread of the social fabric, including the way we experience reality, before we can truly live in harmony with each other and with the natural world.

By the same token, domination preceded exploitation and may long survive an exploitative society if we are not mindful of its scope. Here, too, we must go far beyond the economic areas and economic relationships to search out sensibilities that seek to control all aspects of the world. The complexity of domination is such that even love, when used to manipulate and control, can produce submission as effectively as outright physical coercion. Indeed, self-domination, working through mechanisms like guilt and a variety of socialization techniques, has brought women, the young, and ethnic groups into complicity with their rulers more effectively than explicit methods of control. This again raises the need to go beyond the traditional "isms" structured around self-interest and economic motivations into the deepest recesses of the self: its formation in a cauldron of competition and conflicting interests whereby individuality is identified with domination, self-development with a mentality formed by rivalry, maturity with adaptation to things as they exist, success with acquisition and the sanctity of the bargain.

Social ecology provides the patterning forms to compare and alter the ensembles of hierarchy and domination that afflict us. Its ecological image of animal-plant communities, or what I prefer to call ecocommunities rather than ecosystems (with its bias for systems theory), challenges the notion that hierarchy exists between species. To designate lions as "kings" and ants as "lowly" is meaningless from an ecological viewpoint: ants, in fact, are far less dispensable in recycling an ecocommunity than lions, and words like "kingly" or "lowly" are extrapolations of our own social relationships into the natural world. As to intraspecific "hierarchies," they are so different in kind, so patently unequal to each other, and so transitory where they seem to exist at all, that a sizable volume would be needed to critically examine them with reasonable thoroughness. Suffice it to say that an elephant cow or "matriarch" who seems to lead a herd stands very much at odds with a baboon "patriarch." And a "queen" bee is simply an essential part of a reproductive organ we call a beehive, not a link in an institutionalized dynasty.[5]

An ecological society is more than a society that tries to check the mounting disequilibrium that exists between humanity and the natural world. Reduced to simple technical or political issues, this anemic view of such a society's function degrades the issues raised by an ecological critique and leads to purely technological and instrumental approaches to ecological problems. Social ecology is, first of all, a *sensibility* that includes not only a critique of hierarchy and domination but a reconstructive outlook that advances a participatory concept of "otherness" and a new appreciation of differentiation as a social and biological desideratum. Formalized into certain basic principles, it is also guided by an ethics that emphasizes variety without structuring differences into a hierarchical order. If I were to single out the precepts for such an ethics, I would be obliged to use two words that give it meaning: *participation* and *differentiation*.

Social ecology is largely a philosophy of participation in the broadest sense of the word. In its emphasis on symbiosis as the most important factor in natural evolution, this philosophy sees ecocommunities as participatory communities. The compensatory manner by which animals and plants foster each other's survival, fecundity, and well-being surpasses the emphasis conventional evolutionary theory places on their "competition" with each other—a word that, together with "fitness," is riddled with ambiguities. Competition may accurately describe the workings of our capitalist market, but it does not include the more meaningful principle of complementarity—which, alas, some "natural law" acolytes have decided to call a "law"—that describes the mutualistic interaction of animals and plants.

Similarly, differentiation not only emphasizes the importance of variety for ecological stability but is also the all-important context for the eventual emergence of a nascent freedom in an ecocommunity. Complexity, a product of variety, is a crucial factor in opening alternative evolutionary pathways. The more differentiated the lifeform and the environment in which it exists, the more acute is its overall sensorium, the greater its flexibility, and the more active its participation in its own evolution.

The two concepts cannot be raised without leading to interaction with each other. The greater the differentiation, the wider is the degree of participation in elaborating the world of life. An ecological ethics not only affirms life, it also focuses on the *creativity* of life.

These concepts extend from nature directly into society. They provide us with principles that overcome the dualism between nature and society—not only in theory but also in practice. Looking back in time, we find that the history of society deliciously grades out of the history of life without either being subsumed by the other. Our earliest institutions were based on blood ties, age groups, and gender functions—all biological facts, yet distinctively social in that natural affinities are given structure and stability, cohered by ideologies, and expanded to include seemingly "alien" groups through marital exogamy and the exchange of gifts. In time, the emergence of early cities expands the social bond to a point where people see themselves not only as kin, but also as a common species—a universal *humanitas*. The idea of citizenship, while never completely supplanting the family tie, opens a new community arena and a wide range of human intercourse. A continuum can be traced from the simplest kinds of biological association between human beings to an ever-expansive social arena that fosters participation and, with it, greater differentiation in functions, institutional forms, and individual personalities. Participation unites the biotic ecocommunity with the social ecocommunity by opening new evolutionary possibilities in society and nature. Differentiation yields richer possibilities for the elaboration of these ecocommunities and adds the dimension of freedom, however nascent in nature or explicit in society.

It is at this point that social ecology becomes overtly *political*. Communities normally exist within communities: individual families within tribes, tribes within tribal confederations, confederations that create cities, and cities that enter into confederal relationships with each other. The problem we face is: Where did we leap out of scale to produce state institutions that began to work against participation and also inhibit differentiation? To put this question in more general terms: Where did we go wrong in our history such that we face a crisis of monumental proportions in our relationships with each other and with the natural world?

We clearly leaped out of scale when we formed the nation-state. And it is not only the scale on which we function that has exploded beyond our comprehension and control, but also the deep wound we have inflicted on our own humanity. Ordinary people find it impossible to participate in a nation: they can belong to it but it never belongs to them. The size of the nation-state renders active citizenship impossible, at least on the national level, and it turns politics, conceived as something more than a media spectacle, into a form of statecraft in which the citizen is increasingly disempowered by authoritarian executive agencies, their legislative minions, and an all-encompassing bureaucracy.

That "politics"—a Hellenic term that once meant the management of the *polis*, or municipality, in face-to-face assemblies and publicly controlled councils—is so far removed from our present experience that the word has acquired a sneeringly pejorative meaning need hardly be emphasized. "Politicians" cut shabby figures today; they are the objects of public mistrust and forbearance. The fact that this word was once a *municipal* term, applicable only to the *polis*, has been all but forgotten. The disempowerment of the citizen and the attribution of political activity lead ultimately to the attrition of the self. The real victim of the depoliticization of the people is the ego and human personality, through the transformation of citizens from publicly active human beings into atomized, trivialized "constituencies" who are preoccupied with their individual survival in a world over which they exercise no control. This wound is still hemorrhaging in the "body politic" and threatens to bleed people of all their humanity.

Going back further in time, we can see that we went wrong when the market system broke through the confines that traditional society established to contain it. Traditional societies had a genuine fear of unbridled commerce and the accumulation of wealth. They saw it as anti-social and demonic, a corrosive effluvium of greed and self-aggrandizement that threatened to dissolve long-established ties based on mutual aid and community welfare. This archaic insight has been proven out with a vengeance. Capitalism is a "system" (if such it can be called) that gives rise to the universal reign of limitless buying and selling, indeed of limitless growth and expansion. The reduction of the citizen to a buyer and a seller in the economic realm, not only a "constituent" in the political, carries marketplace rivalry into the most intimate everyday aspects of life. We not only engage in a "struggle" with nature, but we also engage in a struggle with each other. Indeed, our struggle with each other is the source of our struggle with nature, a fact that was to entrap both Marx and one of traditional anarchism's revered "fathers." In Marx's case, this ensemble of struggles gave rise to radical theories in which the emergence of class societies was seen as a desideratum, and the bourgeoisie was cast as a "permanently revolutionary" class in the historic drama of "man's" ascent from "animality."[6] Capitalist ideology goes even further: it not only claims that freedom presupposes the domination of nature, but it sees the domination of nature as an ongoing struggle, a process of *social* selection in which the "fit" survive, no less in society than in nature, while those who cannot "succeed" in both realms fall by the wayside.

For a generation that has experienced the horrors of Auschwitz, the viciousness of this view would hardly require emphasis if it were not incorporated by implication—at times, even explicitly—into "natural-law" theory. Historically, it not only stands at odds with the value systems of all precapitalist societies, which emphasized the virtues of cooperation and giving, but also provides a ghastly apologia for the wounds that have been inflicted on the natural world.

That trade was potentially evil, and profit-making outright sin was a theme that ran through the morality of all precapitalist societies and served in great measure to block the ascendancy of early capitalist relationships over society as a whole. The cultural barriers that precapitalist societies raised against incipient forms of capitalism impeded the latter's development for thousands of years. It was not until the eighteenth century and primarily in England that capitalist market relations finally broke through these barriers and proceeded to spread like an aggressive cancer throughout the world.

My use of the word "cancer" is deliberate and literal, not merely metaphorical. Capitalism, I would argue, is the cancer of society— not simply a *social* cancer, a concept that implies it is some form of human consociation. It is not a social phenomenon but rather an economic one; indeed, it is the substitution of economy for society, the ascendancy of the buyer-seller relationship, mediated by things called "commodities," over the richly articulated social ties that past civilizations, at their best, elaborated and developed for thousands of years in networks of mutual aid, reciprocity, complementarity, and other support systems which made social life meaningful and humanizing. Like all uncontrollable cancers, capitalism has shown that it can grow indefinitely and spread into every social domain that harbored ties of mutuality and collective concern.

If there are any "limits" to the growth of capitalism, they are to be found not in any of its so-called internal contradictions, such as economic breakdown or class wars between the workers and bourgeoisie, as so many radical economists tell us, but in the destruction of that host we call "society," the host that this cancer parasitizes and threatens to annihilate. To respond properly to this crisis, we must develop not only specific antibodies that will arrest the disease, and admittedly the valuable palliatives that will slow up its growth, but also a new immunological system that will make society completely resistant to its recurrence. And I speak, here, of a *system* rather than a contrivance or "technological fix," so strongly favored by the so-called realists in the environmental movement—an ensemble of new sensibilities, cultural forms, a moral economy in which the word "economy" implies a recovery of the original Greek origin of the term as the management of the *oikos*, or household, and a new politics, which recovers its Greek definition as the management of the local community.

The culture and sensibility I call "ecological" is not the primary concern of this book. I have dealt with it in great detail in other works.[7] In any case, they can be inferred quite easily from my remarks on the ethical importance of dealing with the "other" in a complementary manner and the emphasis I place on participation and differentiation as the great motifs in organic evolution. The essays "What Is Social Ecology?" and "Market Economy or Moral Economy?" in fact are permeated by an ethical message that raises cultural and attitudinal solutions to the problems we face. My main concern, however, is to examine the ethical bases, to the extent that these exist in nature and the economy, for an ecological politics. I am deeply concerned with what constitutes the proper domain for that kind of politics and the reconstructive steps that can be taken to remove, in a creative way, the causes that are leading to either ecological breakdown or a nuclear holocaust.

If we rely on self-interest and economic motives to evoke the popular response that will deal with these overarching problems, we will be relying on the very constellation of psychological factors that have so decisively contributed to their emergence. Here we encounter the ironic perversity of a "pragmatism" that is no different, in principle, from the problems it hopes to resolve. Nor can we rely on a politics of media manipulation and party mobilization that really hitches the "masses" to statecraft. Statecraft is for statesmen and the "politics" it generates turns the most dedicated idealists into sleazy politicians. This is not, to be sure, because of bad intentions, but rather because of the exigencies of power-brokering, parliamentarianism, and the inevitable effect of a puffed up, larger-than-life public imagery on ordinary mortals.

Are we obliged, then, to fall back on the good that exists in human nature? It is fashionable, today, to deprecate this element in political life, certainly in a society that assumes that human nature is at best a blank page on which the environment can inscribe anything, or at worst a malignant evil that must be kept in tow by coercion and fear.

Yet let us not sell this factor short. Human beings exhibit more care, dedication, and love than most students of their psyches are willing to acknowledge. There is a great deal of truth to Jules Michelet's description of the French people during the opening months of their great revolution and the sentiment that filled them—a description, in fact, that one often encounters in most hard-nosed, skeptical, and amoral historians of other revolutions as well. "To attain unity," declares the old historian of his country's revolutionary rebirth, "nothing was able to prove an impediment, no surprise was considered too dear. All at once, and without even perceiving it, the [French people] have forgotten the things for which they would have sacrificed their lives the day before, their provincial sentiments, local traditions, and legends. Time and space, those material conditions to which life is subject, are no more. A strange *vita nuova*, one eminently spiritual, and making her revolution a sort of dream, at one time delightful, at another terrible, is now beginning for France. It knew neither time nor space."[8]

Perhaps this is overstated. But there is enough truth in it to say that historical moments do arrive when human beings, collectively as well as singly, exhibit a sense of solidarity, care, and dedication that goes beyond mere inspiration to become a devout passion. We would be hard put to explain how many movements for moral regeneration—for example, Christianity, Islam, Buddhism, or even various forms of socialism—could be reduced to "material interest," a Bakunin or a Marx notwithstanding.[9] The cries of "Life, liberty, and the pursuit of happiness" and "Liberty, Equality, and Fraternity," hollow as they may seem today in the ears of many self-styled radicals, have a magnificent utopian dimension to them, a universal appeal that transcends the various conflicting interests that tried to manipulate the American and French revolutions for their own benefit. Indeed, it is doubtful that millions of people could have been set in motion, often exhibiting a stronger spirit of self-sacrifice than self-interest, if these appeals had spoken only to the economic concerns of calculating merchants and competitive capitalists. Even if one allows for the worst intentions of the few who used the good intentions of the many for their own private ends, the reality of these appeals consists in the empirical fact that the many *did* form the great majority of the people and their intentions made social change possible and tangible, however much it was perverted later on.

If we are to explore human nature, we cannot ignore certain features about it that justify a belief in its cooperative and life-affirming tendencies. Human beings are more helpless and dependent at birth than most animals; their development to maturity requires more time than their nearest primate cousins. This protracted period of development which makes for the mental ability of humans to form a culture also fosters a deep sense of interdependence that promotes the formation and stability of community. We are imminently social animals not because of instinct but rather because we must cooperate with each other to mature in a healthy fashion, not only to survive. This kind of cooperation, which involves a long period of parenting, indeed of human touch, makes for a strong need to associate with others of our own kind. The worst punishment that can be inflicted on any normal human being is isolation, and the most serious emotional trauma the individual seems to suffer is separation. The love, care, aid, and goodwill that a group can furnish to an individual are perhaps the most important contribution it can make to an individual's ego development. Denied these supportive attentions, ego-formation, personal development, and individuality become warped. Speaking in ecological terms, the making of that "whole" we call a rounded, creative, and richly variegated human being crucially depends upon community supports for which no amount of self-interest and egotism is a substitute. Indeed, without these supports, there would be no real self to distort—only a fragmented, wriggling, frail, and pathological thing that could only be called a "self" for want of another word to describe it.

The making of a human being, in short, is a collective process, a process in which both the community and the individual *participate*. It is also a process that, at its best, evokes by its own variety of stimuli the wealth of abilities and traits within the individual that achieve their full degree of *differentiation*. The extent to which these individual potentialities are realized, the unity of diversity they achieve, and the scope they acquire depend crucially upon the degree to which the community itself is participatory and richly differentiated in the stimuli, forms, and choices it creates that make for personal self-formation. Denied the opportunity to participate in a community, whether because it is incomprehensibly large or socially exclusive, the individual begins to feel disempowered and ineffectual, with the result that his or her ego begins to shrivel. Divested of differentiated stimuli, opportunities, choices, and variegated groups that speak to his or her proclivities, the individual becomes a homogenized thing, passive, obedient, and privatized, which makes for a submissive personality and a manipulable constituent.

The principles of social ecology, structured around participation and differentiation, thus reach beyond the biotic ecocommunity directly into the social one, indeed, into the nature of the ego itself and the image it forms of the other. An ecological ethics of freedom thus coheres nature, society, and the individual into a unified whole that leaves the integrity of each untouched and free of a reductionist biologism or an antagonistic dualism. The social derives from the natural and the individual from the social, each retaining its own integrity and specificity through a process of ecological derivation. The great splits between nature and society and between society and individuality are thus healed. They are healed not by any bridge, a term that implies the existence of chasms that are crossed by a structure, but by the very *process* of derivation—that is, by the fact that the individual is the history of individuality as it emerges from society, and society *is* the history of society as it emerges from natural history itself. So, too, is mind in its relationship to body, thought in relationship to physicality, the "I" in relationship to the "other," a liberatory, objective ethics in relationship to the nascent freedom that emerges in the natural world, and humanity in relationship to nature.

The ecological and eminently ethical principles advanced here open distinctly reconstructive avenues in our efforts to resolve the crisis created by ecological breakdown and a world that lives under the shadow of thermonuclear extinction. These avenues are the overarching thoughts in the latter half of my "An Appeal for Social and Ecological Sanity," and they need not be examined in detail here. One focal issue, however, would benefit from some elucidation: the centrality I give to libertarian municipalism or communalism.

Town, city, and neighborhood are the most intimate environments that extend beyond the home and the place of work. For the young, and many women, they are often the only ones that exist. There was a time when the workplace was part of the immediate community and life was lived, as it still is in many places outside megalopolitan areas, in the immediate proximity of a person's household. The medieval town, like so many ancient ones, was intensely peopled; it was the object of firm, personal loyalties, the public sphere in an emotional as well as a political sense, the most important terrain for self-formation beyond the immediate family. Most importantly, it was the arena in which people empowered themselves in assemblies, publicly controlled councils, and in plazas, squares, and other gathering places where they could discuss and resolve public problems.

To be a "citizen" was not a legal abstraction, a juridical void to be filled by rules and regulations delivered by godlike powers far removed from one's personal horizon of the world. Citizenship connoted a high degree of participation, be it in face-to-face decision making or administrative involvement. Politics, in effect, was the notion of community seen as *communizing*, actively interrelating in formal assemblies and informal discussions. To be political meant to be communal, not to be a politician—a creature set aside from the community to be chosen or "elected" for the Elect, who alone could rule and command. These privileged attributes belong to the statesman who engages in the business of "*the* State"—a special apparatus set aside from the community solely to sit at its top with the full weight of authority and control of the means of violence.

By contrast, the municipality formed an arena in its own right. It emerged out of the *social* world in which people engage in their private affairs to develop into a *political* world in which they engage in their public affairs. Historically, it preceded the state with its apparatus of police, soldiers, courts, jails, bureaucracies, and the like. This kind of machinery appeared in cities as well, but usually when they entered into periods of decline and paved the way for the emergence of the state.

If, as I believe, the municipality is increasingly becoming a battleground on which civic politics belligerently confronts state manipulation, this is due in no small measure to the fact that the state, until comparatively recent times, has never been able to *fully* claim the municipality as its own. Like precapitalist societies that blocked the emergence of capitalism with their strong grassroots traditions and entrenched cultural forms, the municipality opposed outright state control with its guilds, neighborhood associations, local societies, and a vast variety of de facto self-governing institutions like the revolutionary sections of Paris in 1793–94 and a host of community organizations in later periods. Municipal life, richly textured by family networks and popular organizations—many of which cut across class lines—has always been a human refuge from the homogenizing and dehumanizing effects of state bureaucracies. This inner cultural strength has made it the bulwark *par excellence* against the encroachment of the state on public life, not only today but also historically.[10]

If the state today, owing in great part to the expansion of the market economy, threatens to destroy this refuge, this means that the municipality is not only faced with the loss of its traditional identity, but also is becoming, by the sheer pressure of events, the most significant terrain for the struggle against the state. Historically, there is nothing new about this confrontation. Almost every major revolution has involved—indeed has often been—a conflict between the local community and the centralized state. And just as the centralized state means the nation-state, so the local community means the municipality—be it the village, town, city, or neighborhood. From the peasant wars in Germany during the 1520s, through the English, American, and French revolutions, including Parisian uprisings from the 1790s to 1871, what we see are local communities pitted against centralized state institutions—a persistent conflict that has yet to receive the attention it deserves. The demand for "local control" does not necessarily mean the parochialism and insularity that evoked so much opposition in Marx's writings. In the force field generated by an increasingly centralized state and increasingly resistant communities, the cry for greater municipal autonomy echoes demands for a new political culture marked by autonomy, relative self-sufficiency, and more open democratic institutions.

To speak in a more constructive vein, the municipality may well be the one arena in which traditional institutional forms can be reworked to replace the nation-state itself. The potential for a truly liberatory radicalism has always been inherent in the municipality; it forms the bedrock for direct political relationships, face-to-face democracy, and new forms of self-governance by neighborhoods and towns. To be sure, the municipality's capacity to play a historic role in changing society today depends on the extent to which it can shake off the state institutions that have infiltrated it: its mayoralty structure, civic bureaucracy, and its own professionalized monopoly on violence. Rescued from these institutions, however, it retains the historic materials and political culture that can pit it against the nation-state and the cancerous corporate world that threatens to digest social life as such.

Let us not deceive ourselves, however, in thinking that a libertarian municipal alternative to the nation-state is meaningful in one or only a few communities. Freedom is not achievable in a lasting form on the margins or in the pockets of society. Left to themselves, state institutions are much too powerful to permit isolated towns and cities to regain their political autonomy. The creativity of municipal politics will ultimately be tested when villages, towns, and cities manage to confederate with each other and form radically new social networks, perhaps on a county level to begin with, later on a regional, and ultimately on a nationwide level.

The possibility that authoritarian forms of coordination will emerge from free municipalities cannot be discounted or legislated away by mere goodwill and idealistic rhetoric. Only insofar as the coordination of municipalities is strictly *administrative* and effectuated by recallable, rotatable, and clearly mandated *deputies* of the people (not their "representatives"), drawn from the citizens assemblies of their own municipalities, can we say that it is structured along libertarian lines. *Policy*, in turn, would have to be the exclusive province of the assemblies, not of elected "legislators." Here,

Rousseau's famous remarks about "representation" are as valid today as they were two centuries ago:

> Sovereignty, for the same reason as it makes it inalienable, cannot be represented. It lies essentially in the general will, and will does not admit of representation: it is either the same, or other; there is no intermediate possibility. The deputies of the people, therefore, are not and cannot be its representatives: they are merely its stewards, and can carry through no definitive acts. Every law the people has not ratified in person is null and void—is, in fact, not a law. The people of England regards itself as free: but it is grossly mistaken: it is free only during the election of members of parliament. As soon as they are elected, slavery overtakes it, and it is nothing.[11]

Viewed philosophically, the free municipality transforms an ecological ethics from the realm of precept into the realm of politics. The Greeks tried to do this in real life when Athens was conceived as an ethical compact between its citizens, not merely as a dwelling place. Social ecology, which tries to plant its feet in nature, begins to raise its head in the municipality that is truly participatory and fosters differentiation—in short, that is truly libertarian. Hence the very natural processes that operate in animal and plant evolution along the symbiotic lines of participation and differentiation reappear as social processes in human evolution, albeit with their own distinctive traits, qualities, and gradations or phases of development. Coherence takes the form not of a mystical teleology that predetermines the end in the beginning of a process, but of a tendency that is unified by the shared history that society has as a result of its emergence from nature, and individuality has as a result of its emergence from society. What would be truly mystical is the notion that social history, which has its ground in natural history, is so severed from its own parentage that nature no longer operates as a basic factor in social development or, for that matter, that an individual's biography is so autonomous that society no longer operates as a basic factor in personal development.

The municipality is close at hand, existential, and ever-present in our lives. The nation-state is remote, largely the product of ideology, and almost ethereal in the ordinary person's experience—except when it invades his or her personal environment with its demands. Our nationality tends to be a media event and our state capitals tourist traps. When we return home from them, we are restored not only to a personal world which we call our "homes," but also a village, town, neighborhood, or city that is the real locus of our lives as social and political beings. The rest is largely synthetic and more contrived than spontaneous. To reflect on these realities of our lives is to break through the fog of nationalist obscurantism and recover not only our sense of place but also our sense of politics. Ecological politics, in this sense, is a politics of *oikos* and community, the eco-community in which people live out their social and political lives in a fully existential and ethically meaningful way.

That this environment must recover its human scale, that it must be decentralized sufficiently so that we can understand it and participate in it, is too obvious to belabor. That it must be freed of those statist and economic constraints which inhibit its spontaneous differentiation into a world rich in its diversity of stimuli, the freedom to create, and the opportunity to choose alternatives that make freedom existentially meaningful is equally obvious by now. That citizenship is an ethical compact, not a commercial contract, is a historic truth to which we must repair if we are to be truly human.

The nation-state makes us less than human. It towers over us, cajoles us, disempowers us, bilks us of our substance, humiliates us—and often kills us in its imperial adventures. To be a citizen of a nation-state is an abstraction which removes us from our lived space to a realm of myth, clothed in the superstition of a "uniqueness" that sets us apart as a national entity, from the rest of humanity—indeed, from our very species. In reality, we are the nation-state's victims, not its constituents—not only physically and psychologically, but also ideologically.

Nothing reveals this more vividly than the extent to which the nation-state has absorbed the energy and belief systems of domestic radical movements. Nearly all of them today have joined the conflict between nation-states and have virtually abandoned their universalist claims to seek human liberation as such. Today they are part of the Cold War and their efforts are deployed in conflicts between "superpowers" and their junior powers in the Third World. Socialism in great part has become a form of national socialism in its quest for national affiliations—its attempt to clothe its ideals or even fit them to meet territorial tangibility and national identity. The deeply humanistic internationalism, indeed *anti*nationalism, that characterized radical movements in the early decades of the twentieth century, and was voiced by revolutionary heroines like Louise Michel and Rosa Luxemburg, has been transformed into a crude nationalism that shrilly participates in imperial designs, even as it professes to oppose imperialism as such. Hence the cynical selectivity of the dogmatic "Left": embarrassment with the Russian invasion of Hungary in 1956; virtual silence about the Russian invasion of Czechoslovakia in 1968; and faint motions of "protest" against the Russian crushing of Polish Solidarity.[12]

The color of radicalism today is no longer red; it is green, and should be raised aloft boldly if the modern crisis is to be resolved. The politics we must pursue is grassroots, fertilized by the ecological, feminist, communitarian, and antiwar movements that have patently displaced the traditional workers' movements of half a century ago. The ethics we need is predicated on a definition of the good, not on calculations of "lesser evils" or "benefits versus risks" that betray us to the worst of evils and the greatest of risks that lie at the end of the road. And the function of our politics must not only be to mobilize, but also to educate, to use knowledge for the empowerment of people, not for their manipulation.

If a green perspective structured around social ecology and its evolutionary vision of freedom does emerge, would it be too bold to say that it will bring together all the threads of the seemingly fragmented development of past decades—a development toward the most expansive and coherent expression of liberation we have known up to now? Would it be reasonable to suppose that the civil rights movement, the counterculture, and the New Left of the early and mid-sixties were the soil for the growth of feminism and gay liberation in the latter half of the decade, for environmentalism and later ecologism in the early seventies, for the persisting communitarian and localist movements in both decades that nourished the anti-nuclear, peace, climate justice, and citizen activism of more recent times—each forming an aspect of a common development with shared roots and expressive of richer phases in the definition and struggle for freedom? The social movements that have marked the past decades serve as a rich continuum that has brought out in ever greater fullness the potentiality for freedom that is latent in our era with all its varied and rich articulations. In any case, each such articulation—be it feminist or peace-oriented, countercultural or environmental, communitarian or localist—remains vibrantly structured in the other and exists as part of a whole that can be regarded as a "new social movement," to use the language of the sociologists, not merely a collection of separate movements that academics inventory in their shopping lists of "new causes" or "failed causes."

Much depends on the level of consciousness such a green movement attains. If it confines itself to evocations of the "simple life," with a biocentricity that ignores humanity's own unique potentialities, an "anti-humanism" that denies what we *can* be as human beings in the larger world of life, or an antirationalism that ignores the organic nature of dialectical reason because it fears the narrow analytic and instrumental reason so prevalent today, then the continuum of the past decades will be broken and this great development of the time in all its phases will be aborted. With such simple intellectual equipment, it will indeed become too simple-minded to be credible and meaningful. In the climate of cooptation and fragmentation of ideas that makes it possible for utterly contradictory ideas and values to exist in the minds of the same individuals as well as the same written works, a due regard for depth and coherence is more necessary today than it has ever been. It is only fair to ask of everyone that she or he *derive* ideas, not merely *collect* them—that there be an explanation of the origins, meaning, development, and direction of ideas, not merely that they be held together with glue and scotch tape.

Finally, if we plan to speak about public issues to the public, we would do well to draw our language from the political culture of the public—not from languages and traditions that are utterly alien to that culture. In America, this political culture stems from a splendid revolutionary tradition marked by a strong libertarian ambience: a reverence for the rights of the individual over those of the state, of the locality over centralized power, of autonomy over dependency, and of self-sufficiency over corporate control. That reactionary movements have coopted these allegiances is evidence not of their reactionary nature, but of the inability of centralistic forms of socialism, with their emphases on centralism and state controls, to address them in an authentically libertarian way. The success of regressive ideologies is often searing evidence of the failings that burden their self-styled "progressive" counterparts.

An ideological vacuum exists in modern society, and its crisis still persists. It will not be for want of solutions that this condition will remain, but rather for want of the willingness to see what has changed in recent decades that renders traditional "isms" obsolete. The answers are gestating in our body politic; what we lack are the obstetricians who can bring them to birth and the educators who can bring them to maturity.

September, 1985

What is Social Ecology?

Social ecology is based on the conviction that nearly all of our present ecological problems originate in deep-seated social problems. It follows, from this view, that these ecological problems cannot be understood, let alone solved, without a careful understanding of our existing society and the irrationalities that dominate it. To make this point more concrete: economic, ethnic, cultural, and gender conflicts, among many others, lie at the core of the most serious ecological dislocations we face today—apart, to be sure, from those that are produced by natural catastrophes. The massive oil spills that have occurred over the past two decades, the extensive deforestation of tropical forest and magnificent ancient trees in temperate areas, and vast hydroelectric projects that flood places where people live, to cite only a few problems, are sobering reminders that the real battleground on which the ecological future of the planet will be decided is clearly a social one, particularly between corporate power and the long-range interests of humanity as a whole.

Indeed, to separate ecological problems from social problems—or even to play down or give only token recognition to their crucial relationship—would be to grossly misconstrue the sources of the growing environmental crisis. In effect, the way human beings deal with each other as social beings is crucial to addressing the ecological crisis. Unless we clearly recognize this, we will fail to see that the hierarchical mentality and class relationships that so thoroughly permeate society are what has given rise to the very idea of dominating the natural world.

Unless we realize that the present market society, structured around the brutally competitive imperative of "grow or die," is a thoroughly impersonal, self-operating mechanism, we will falsely tend to blame other phenomena—such as technology or population growth—for growing environmental dislocations. We will ignore their *root* causes, such as trade for profit, industrial expansion for its own sake, and the identification of progress with corporate self-interest. In short, we will tend to focus on the *symptoms* of a grim social pathology rather than on the pathology itself, and our efforts will be directed toward limited goals whose attainment is more cosmetic than curative.

Nature and Society

To escape from this profit-oriented image of ecology, let us begin with some basics—namely, by asking what society and the natural world actually are. Among the many definitions of *nature* that have been formulated over time, the one that has the most affinity with social ecology is rather elusive and often difficult to grasp because understanding and articulating it requires a certain way of thinking—one that stands *at odds* with what is popularly called "linear thinking." This "nonlinear" or organic way of thinking is developmental rather than analytical, or in more technical terms, it is dialectical rather than instrumental. It conceives the natural world as a *developmental process*, rather than the beautiful vistas we see from a mountaintop or images fixed on the backs of picture postcards. Such vistas and images of nonhuman nature are basically static and immobile. As we gaze over a landscape, to be sure, our attention may momentarily be arrested by the soaring flight of a hawk, or the bolting leap of a deer, or the low-slung shadowy lope of a coyote. But what we are really witnessing in such cases is the mere kinetics of physical motion, caught in the frame of an essentially static image of the scene before our eyes. Such static images deceive us into believing in the "eternality" of single moments in nature.

But nonhuman nature is more than a scenic view, and as we examine it with some care, we begin to sense that it is basically an evolving and unfolding phenomenon, a richly fecund, even dramatic development that is forever changing. I mean to define nonhuman nature precisely as an evolving process, as the *totality*, in fact, of its evolution. Nature, so concerned, encompasses the development from the inorganic into the organic, and from the less differentiated and relatively limited world of unicellular organisms into that of multicellular ones equipped with simple, then, complex, and in time fairly intelligent neural apparatuses that allow them to make innovative choices. Finally, the acquisition of warm-bloodedness gives to organisms the astonishing flexibility to exist in the most demanding climatic environments.

This vast drama of nonhuman nature is in every respect stunning and wondrous. Its evolution is marked by increasing subjectivity and flexibility and by increasing differentiation that makes an organism more adaptable to new environmental challenges and opportunities and that better equips living beings (specifically human beings) to *alter* their environment to meet their own needs rather than merely adapt to environmental changes. One may speculate that the potentiality of matter itself—the ceaseless interactivity of atoms in forming new chemical combinations to produce ever more complex molecules, amino acids, proteins, and under suitable conditions, elementary life-forms—is inherent in inorganic nature.[1] Or one may decide quite matter-of-factly that the "struggle for existence" or the "survival of the fittest" explains why increasingly subjective and more flexible beings are capable of addressing environmental change more effectively than are less subjective and flexible beings. But the simple fact remains that these evolutionary dramas did occur, indeed the evidence is carved in stone in the fossil record. That nonhuman nature is this record, this history, this developmental or evolutionary process, is a very sobering fact that cannot be ignored without ignoring reality itself.

Conceiving nonhuman nature as its own interactive evolution rather than as a mere scenic vista has profound implications—ethical as well as biological—for ecologically minded people. Human beings embody, at least potentially, attributes of nonhuman development that place them squarely within organic evolution. They are not "natural aliens," to use Neil Evernden's phrase, strong exotics, phylogenetic deformities that, owing to their tool-making capacities, "cannot evolve *with* an ecosystem anywhere."² Nor are they "intelligent fleas," to use the language of Gaian theorists who believe that the earth ("Gaia") is one living organism.³ These untenable disjunctions between humanity and the evolutionary process are as superficial as they are potentially misanthropic. Humans are highly intelligent, indeed, very self-conscious primates, which is to say that they have emerged—not diverged—from a long evolution of vertebrate life-forms into mammalian and finally primate life-forms. They are a product of a significant evolutionary trend toward intellectuality, self-awareness, will, intentionality, and expressiveness, be it in verbal or in body language.

Human beings belong to a natural continuum, no less than their primate ancestors and mammals in general. To depict them as "aliens" that have no place or pedigree in natural evolution, or to see them essentially as an infestation that parasitizes the planet the way fleas parasitize dogs and cats, is not only bad ecology but bad thinking. Lacking any sense of process, this kind of thinking—regrettably so commonplace among ethicists—radically divides the nonhuman from the human. Indeed, to the degree environmental thinkers romanticize nonhuman nature as wilderness and see it as more authentically "natural" than the works of humans, they freeze nonhuman nature as a circumscribed domain in which human innovation, foresight, and creativity have no place and offer no possibilities.

The truth is that human beings not only belong in nature, they are products of a long, natural evolutionary process. Their seemingly "unnatural" activities—like the development of technology and science, the formation of mutable social institutions, highly symbolic forms of communication, and aesthetic sensibilities, and the creation of towns and cities—all would have been impossible without the large array of physical human attributes that have been aeons in the making, be they the large human brain or the bipedal motion that frees human hands for making tools and carrying food. In many respects, human traits are enlargements of nonhuman traits that have been evolving over the ages. Increasing care for the young, cooperation, the substitution of mentally guided behavior for largely instinctive behavior—all are present more keenly in human behavior. Among humans, as opposed to nonhuman beings, these traits are developed sufficiently to reach a degree of elaboration and integration that yields cultures, comprising institutions of families, bands, tribes, hierarchies, economic classes, and the state—in short, highly mutable *societies* for which there is no precedent in the nonhuman world, unless the genetically programmed behavior of insects is to be regarded as social. In fact, the emergence and development of human society has been a continual process of shedding instinctive behavioral traits and of clearing a new terrain for potentially rational behavior.

Human beings always remain rooted in their biological evolutionary history, which we may call "first nature," but they produce a characteristically human social nature of their own, which we may call "second nature." Far from being unnatural, human second nature is eminently a creation of organic evolution's first nature. To write second nature out of nature as a whole, or indeed to minimize it, is to ignore the creativity of natural evolution itself and to view it one-sidedly. If "true" evolution embodies itself simply in creatures like grizzly bears, wolves, and whales—generally, animals that *people* find aesthetically pleasing or relatively intelligent—then human beings are *de*-natured. Such views, whether they see human beings as "aliens" or as "fleas," essentially place them outside the self-organizing thrust of natural evolution toward increasing subjectivity and flexibility. The more enthusiastic proponents of this de-naturing of humanity may see human beings as existing apart from nonhuman evolution, as a "freaking," as Paul Shepard put it, of the evolutionary process. Others simply avoid the problem of clarifying humanity's unique place in natural evolution by promiscuously putting human beings on a par with beetles in terms of their "intrinsic worth." The "either/or" propositional thinking that produces such obfuscations either separates the social from the organic altogether or flippantly makes it disappear into the organic, resulting in an inexplicable dualism at one extreme or a naïve reductionism at the other. The dualistic approach, with its quasi-theological premise that the world was "made" for human use, is saddled with the name *anthropocentrism*, while the reductionist approach, with its almost meaningless notion of a "biocentric democracy," is saddled with the name *biocentrism*.

The bifurcation of the human from the nonhuman reflects a failure to think organically or to approach evolutionary phenomena with an evolutionary way of thought. Needless to say, if nature were no more than a scenic vista, then mere metaphoric and poetic descriptions of it might suffice to replace systematic thinking about it. But *nature is the history of nature*, an evolutionary process that is going on to one degree or another under our very eyes, and as such, we dishonor it by thinking of it in anything but a processual way. That is to say, we require a way of thinking that recognizes that "what is," as it seems to lie before our eyes, is always developing into "what is not," that it is engaged in a continual self-organizing process in which past and present, along a richly differentiated but shared continuum, give rise to a new potentiality for an ever-richer degree of *wholeness*. Life, clearly in its human form, becomes open-endedly innovative and transcends its relatively narrow capacity to adapt only to a pregiven set of environmental conditions. As V. Gordon Childe once put it, "Man makes himself; he is not preset to survive by his genetic makeup."

By the same token, a processual, organic, and dialectical way of thinking has little difficulty in locating and explaining the emergence of the social out of the biological, of second nature out of first nature. To truly *know* and be able to give interpretive *meaning* to the social issues and ideas so arranged, we should want to know how each one derived from the other and what its part is in an overall development. What, in fact, is meant by "decentralization," and how, in the history of human society, does it derive from or give rise to centralization? We need processual thinking to comprehend processual realities, if we are to gain some sense of *direction*—practical as well as theoretical—in addressing our ecological problems.

Social ecology seems to stand alone, at present, in calling for an organic, developmental way of thinking out problems that are basically organic and developmental in character. The very definition of the natural world *as* a development indicates the need for organic thinking, as does the derivation of human from nonhuman nature—a derivation from which we can draw far-reaching conclusions for the development of an ecological ethics that in turn can provide serious guidelines for the solution of our ecological problems.

Social ecology calls upon us to see that the natural world and the social are interlinked by evolution into one nature that consists of two differentiations: first or biotic nature, and second or social nature. Social nature and biotic nature share an evolutionary potential for greater subjectivity and flexibility. Second nature is the way in which human beings, as flexible, highly intelligent primates, inhabit and *alter* the natural world. That is to say, people create an environment that is most suitable for their mode of existence. In this respect, second nature is no different from the environment that *every* animal, depending upon its abilities, partially creates as well as primarily adapts to—the biophysical circumstances or ecocommunity in which it must live. In principle, on this very simple level, human beings are doing nothing that differs from the survival activities of nonhuman beings, be it building beaver dams or digging gopher holes.

But the environmental changes that human beings produce are profoundly different from those produced by nonhuman beings. Humans act upon their environments with considerable technical *foresight*, however lacking that foresight may be in ecological ideals. Animals adapt to the world around them; human beings innovate through thought and social labor. For better or worse, they alter the natural world to meet their needs and desires—not because they are perverse, but because they have evolved quite naturally over the ages to do so. Their cultures are rich in knowledge, experience, cooperation, and conceptual intellectuality; however, they have been sharply divided against themselves at many points of their development, through conflicts between groups, classes, nation-states, and even city-states. Nonhuman beings generally live in ecological niches, their behavior guided primarily by instinctive drives and conditioned reflexes. Human societies are "bonded" together by *institutions* that change radically over centuries. Nonhuman communities are

notable for their general fixity, by their clearly preset, often genetically imprinted rhythms. Human communities are guided in part
by ideological factors and are subject to changes conditioned by
those factors. Nonhuman communities are generally tied together
by genetically rooted instinctive factors—to the extent that these
communities exist at all.

Hence human beings, emerging from an organic evolutionary
process, initiate, by the sheer force of their biological and survival
needs, a social evolutionary development that clearly involves their
organic evolutionary process. Owing to their naturally endowed
intelligence, powers of communication, capacity for institutional
organization, and relative freedom from instinctive behavior, they
refashion their environment—as do nonhuman beings—to the full
extent that their biological equipment allows. This equipment makes
it possible for them to engage not only in social life but in social
development. It is not so much that human beings, in principle,
behave differently from animals or are inherently more problematical in a strictly ecological sense, as it is that the social development
by which they grade out of their biological development often
becomes more problematical for themselves and nonhuman life.
How these problems emerge, the ideologies they produce, the extent
to which they contribute to biotic evolution or abort it, and the damage they inflict on the planet as a whole lie at the very heart of the
modern ecological crisis. Second nature as it exists today, far from
marking the fulfillment of human potentialities, is riddled by contradictions, antagonisms, and conflicting interests that have distorted
humanity's unique capacities for development. Its future prospects
encompass both the danger of tearing down the biosphere and, given
the struggle to achieve an ecological society, the capacity to provide
an entirely new ecological dispensation.

Social Hierarchy and Domination

How, then, did the social emerge from the biological? We have good
reason to believe that as biological facts such as kin lineage, gender

distinctions, and age differences were slowly institutionalized, their uniquely social dimension was initially quite egalitarian. Later this development acquired an oppressive hierarchical and then an exploitative class form. The lineage or blood tie in early prehistory obviously formed the organic basis of the family. Indeed, it joined together groups of families into bands, clans, and tribes, through either intermarriage or fictive forms of descent, thereby forming the earliest social horizon of our ancestors. More than in other mammals, the simple biological facts of human reproduction and the protracted maternal care of the human infant tended to knit siblings together and produced a strong sense of solidarity and group inwardness. Men, women, and their children were socialized by means of a fairly stable family life, based on mutual obligation and an expressed affinity that was often sanctified by initiation ceremonies and marital vows of one kind or another.

Human beings who were outside the family and all its elaborations into bands, clans, tribes, and the like, were regarded as "strangers" who could alternatively be welcomed hospitably or enslaved or put to death. What mores existed were based on unreflective *customs* that seemed to have been inherited from time immemorial. What we call *morality* began as the rules or commandments of a deity or various deities, in that moral beliefs required some kind of supernatural or mystical reinforcement or sanctification to be accepted by a community. Only later, beginning with the ancient Greeks, did *ethics* emerge, based on rational discourse and reflection. The shift from blind custom to a commanding morality and finally to a rational ethics occurred with the rise of cities and urban cosmopolitanism, although by no means did custom and morality diminish in importance. Humanity, gradually disengaging its social organization from the biological facts of blood ties, began to admit the "stranger" and increasingly recognize itself as a shared community of human beings (and ultimately a community of citizens) rather than an ethnic folk or group of kinsmen.

In this primordial and socially formative world, other human biological traits were also reworked from the strictly natural to the social. One of these was the fact of age and its distinctions. In social groups among early humans, the absence of a written language helped to confer on the elderly a high degree of status, for it was they who possessed the traditional wisdom of the community, including knowledge of the traditional kinship lines that prescribed marital ties in obedience to extensive incest taboos as well as survival techniques that had to be acquired by both the young and the mature members of the group. In addition, the *biological* fact of gender distinctions was slowly reworked along *social* lines into what were initially complementary sororal and fraternal groups. Women formed their own food-gathering and care-taking groups with their own customs, belief systems, and values, while men formed their own hunting and warrior groups with their own behavioral characteristics, mores, and ideologies.

From everything we know about the socialization of the biological facts of kinship, age, and gender groups—their elaboration into early institutions—there is no reason to doubt that these groups existed initially in complementary relationships with one another. Each, in effect, needed the others to form a relatively stable whole. No one group "dominated" the others or tried to privilege itself in the normal course of things. Yet even as the biological underpinnings of consociation were, over time, further reworked into social institutions, so the social institutions were slowly reworked, at various periods and in various degrees, into hierarchical structures based on command and obedience. I speak here of a historical trend, in no way predetermined by any mystical force or deity, and one that was often a very limited development among many preliterate or aboriginal cultures and even in certain fairly elaborate civilizations.

Hierarchy in its earliest forms was probably not marked by the harsh qualities it has acquired over history. Elders, at the very beginnings of gerontocracy, were not only respected for their wisdom but were often beloved of the young, with affection that was often reciprocated in kind. We can probably account for the increasing harshness of later gerontocracies by supposing that the elderly, burdened by their failing physical powers and dependent upon their community's goodwill, were more vulnerable to abandonment in periods of material want than any other part of the population. "Even in simple food-gathering cultures," observed anthropologist Paul Radin, "individuals above fifty, let us say, apparently arrogate to themselves certain powers and privileges which benefited themselves specifically, and were not necessarily, if at all, dictated by considerations either of the rights of others or the welfare of the community."[4] In any case, that gerontocracy was probably the earliest form of hierarchy is corroborated by its existence in communities as disparate as the Australian Aborigines, tribal societies in East Africa, and Native communities in the Americas. Many tribal councils throughout the world were really councils of elders, an institution that never completely disappeared (as the word *alderman* suggests), even after they were overlaid by warrior societies, chiefdoms, and kingships.

Patricentricity, in which masculine values, institutions, and forms of behavior prevail over feminine ones, seems to have developed in the wake of gerontocracy. Initially, the emergence of patricentricity may have been a useful adjunct to a life deeply rooted in the primordial natural world; preliterate and early aboriginal societies were essentially small domestic communities in which the authentic center of material life was the home, not the "men's house" so widely present in later, more elaborate tribal societies. Male rule, if such it can strictly be called, takes on its harshest and most coercive form in *patriarchy*, an institution in which the eldest male of an extended family or clan has a life-and-death command over *all* other members of the group. Women may be ordered whom to marry, but they are by no means the exclusive or even the principal object of a patriarch's domination. Sons, like daughters, may be ordered how to behave at the patriarch's command or be killed at his whim.

So far as patricentricity is concerned, however, the authority and prerogative of the male are the product of a long, often subtly negotiated development in which the male fraternity edges out the female sorority by virtue of the former's growing "civil" responsibilities. Increasing population, marauding bands of outsiders whose migrations may be induced by drought or other unfavorable conditions, and vendettas of one kind or another, to cite common causes of hostility or war, create a new "civil" sphere side by side with woman's domestic sphere, and the former gradually encroaches upon the latter. With the appearance of cattle-drawn plow agriculture, the male, who is the "master of the beasts," begins to invade the horticultural sphere of woman, whose primacy as the food cultivator and food gatherer gives her cultural preeminence in the community's internal life, slowly diluting her preeminence. Warrior societies and chiefdoms carry the momentum of male dominance to the level of a new material and cultural dispensation. Male dominance becomes extremely active and ultimately yields a world in which male elites dominate not only women but also, in the form of classes, other men.

The causes of the emergence of hierarchy are transparent enough: the infirmities of age, increasing population numbers, natural disasters, technological changes that privileged activities of hunting and animal husbandry over horticultural responsibilities, the growth of civil society, and the spread of warfare, all served to enhance the male's standing at the expense of the female's. It must be emphasized that hierarchical domination, however coercive it may be, is not the same thing as class exploitation. As I wrote in *The Ecology of Freedom*,

> Hierarchy must be viewed as *institutionalized* relationships, relationships that living beings literally institute or create but which are neither ruthlessly fixed by instinct on the one hand nor idiosyncratic on the other. By this, I mean that they must comprise a clearly *social* structure of coercive and privileged ranks that exist apart from the idiosyncratic individuals who seem to be dominant within a given community, a hierarchy that is guided by a social logic that goes beyond individual interactions or inborn patterns of behavior.[5]

Marxist theorists tend to single out technological advances and the presumed material surpluses they produce to explain the emergence of elite strata—indeed, of exploiting ruling classes. However, this does not tell us why many societies whose environments were abundantly rich in food never produced such strata. That surpluses are necessary to support elites and classes is obvious, as Aristotle pointed out more than two millennia ago, but too many communities that had such resources at their disposal remained quite egalitarian and never "advanced" to hierarchical or class societies.

It is worth emphasizing that hierarchical domination, however coercive it may be, is not to be confused with class exploitation. Often the role of high-status individuals is very well-meaning, as in the case of commands given by caring parents to their children, of concerned husbands and wives to each other, or of elderly people to younger ones. In tribal societies, even where a considerable measure of authority accrues to a chief—and most chiefs are advisers rather than rulers—he usually must earn the esteem of the community by interacting with the people, and he can easily be ignored or removed from his position by them. Many chiefs earn their prestige, so essential to their authority, by disposing of gifts, and even by a considerable disaccumulation of their personal goods. The respect accorded to many chiefs is earned, not by hoarding surpluses as a means to power but by disposing of them as evidence of generosity.

By contrast, classes tend to operate along different lines. In class societies power is usually gained by the *acquisition* of wealth, not by its disposal; rulership is guaranteed by outright physical coercion, not simply by persuasion; and the state is the ultimate guarantor of authority. That hierarchy is historically more entrenched than class can perhaps be verified by the fact that, despite sweeping changes in class societies, even of an economically egalitarian kind, women have still been dominated beings for millennia. By the same token, the abolition of class rule and economic exploitation offers no guarantee whatever that elaborate hierarchies and systems of domination will also disappear.

In nonhierarchical societies, certain customs guide human behavior along basically decent lines. Of primary importance among early customs was the principle of the *irreducible minimum* (to use Paul Radin's expression), the shared notion that all members of the same community are entitled to the means of life, irrespective of the amount of work they perform. To deny anyone food, shelter, and the basic means of life because of their infirmities or even their frivolous behavior would have been seen as a heinous denial of the very right to live. Nor were the basic resources needed to sustain the community ever permitted to be privately owned; overriding individualistic control was the broader principle of *usufruct*—the notion that the means of life that were not being used by one group could be used, as needed, by another. Thus, unused land, orchards, and even tools and weapons, if left idle, were often at the disposition of anyone in the community who needed them. Lastly, custom fostered the practice of *mutual aid*, the rather sensible cooperative sharing of things and labor, so that an individual or family in straitened circumstances could expect to be helped by others. Taken as whole, these customs became so sedimented into organic society that they persisted long after hierarchy became oppressive and class society became predominant.

The Idea of Dominating Nature

Nature, in the sense of the biotic environment from which humans take the simple things they need for survival, often has no meaning to preliterate peoples as a general concept. Immersed in it as they are, even celebrating animistic rituals in an environment they view as a nexus of life, often imputing their own social institutions to the behavior of nonhuman species, as in the case of beaver "lodges" and humanlike spirits, the concept of "nature" as such eludes them. Words that express our conventional notions of nature are not easy to find, if they exist at all, in the languages of aboriginal peoples.

With the rise of hierarchy and domination, however, the seeds were planted for the belief that first nature not only exists as a world that is increasingly distinguishable from the community but one that is hierarchically organized and can be dominated by human beings. The worldview of magic reveals this shift clearly. Here nature was not conceived as a world apart; rather, a practitioner of magic essentially pleaded with the "chief spirit" of a game animal (itself a puzzling figure in the dream world) to coax it in the direction of an arrow or a spear. Later, magic became almost entirely instrumental; the hunter used magical techniques to "coerce" the game to become prey. While the earliest forms of magic may be regarded as the practices of a generally nonhierarchical and egalitarian community, the later kinds of animistic beliefs betray a more or less hierarchical view of the natural world and of latent human powers of domination over reality.

We must emphasize here that the *idea* of dominating nature has its primary source in the domination of human by human and in the structuring of the natural world into a hierarchical chain of being (a static conception, incidentally, that has no relationship to the dynamic evolution of life into increasingly advanced forms of subjectivity and flexibility). The biblical injunction that gave command of the living world to Adam and Noah was above all an expression of a *social* dispensation. Its idea of dominating nature—so essential to the view of the nonhuman world as an object of domination—can be overcome only through the creation of a society without those class and hierarchical structures that make for rule and obedience in private as well as public life, and the objectifications of reality as mere materials for exploitation. That this revolutionary dispensation would involve changes in attitudes and values should go without saying. But new ecological attitudes and values will remain vaporous if they are not given substance and solidity through real and objective institutions (the structures by which humans concretely interact with each other) and through the tangible realities of everyday life from childrearing to work and play. Until human beings cease to live in societies that are structured around hierarchies as well as economic classes, we shall never be free of domination, however much we may try to dispel it with rituals, incantations, ecotheologies, and the adoption of seemingly "natural" lifeways.[6]

The idea of dominating nature has a history that is almost as old as that of hierarchy itself. Already in the *Gilgamesh* epic of Mesopotamia, a drama whose written form dates back some four thousand years, the hero defies the deities and cuts down their sacred trees in his quest for immortality. The *Odyssey* is a vast travelogue of the Greek warrior, more canny than heroic, who in his wanderings essentially subdues the nature deities that the Hellenic world had inherited from its less well-known precursors (ironically, the dark pre-Olympian world that has been revived by purveyors of eco-mysticism and spiritualism). Long before the emergence of modern science, "linear" rationality, and "industrial society" (to cite causal factors that are often invoked flippantly by some in the modern ecology movement), hierarchical and class societies laid waste to much of the Mediterranean basin as well as the hillsides of China, beginning a vast remaking and often despoliation of the planet.

To be sure, human second nature, in inflicting harm on first nature, created no Garden of Eden. More often than not, it despoiled much that was beautiful, creative, and dynamic in the biotic world, just as it ravaged human life itself in murderous warfare, genocide, and acts of heartless oppression. Social ecology maintains that the future of human life goes hand in hand with the future of the non-human world, yet it does not overlook the fact that the harm that hierarchical and class society inflicted on the natural world was more than matched by the harm it inflicted on much of humanity.

However troubling the ills produced by second nature, the customs of the irreducible minimum, usufruct, and mutual aid cannot be ignored in any account of anthropology and history. These customs persisted well into historical times and surfaced sometimes explosively in massive popular uprisings, from revolts in ancient Sumer to the present time. Many of those revolts demanded the recovery of caring and communistic values, at times when these were coming under the onslaught of elitist and class oppression. Indeed, despite the armies that roamed the landscape of warring areas, the tax-gatherers who plundered ordinary village peoples, and the daily abuses that overseers inflicted on peasants and workers, community life still persisted and retained many of the cherished values of a more egalitarian past. Neither ancient despots nor feudal lords could fully obliterate them in peasant villages and in the towns with independent craft associations. In ancient Greece, a rational philosophy that rejected the encumbering of thought and political life by extravagant wants, as well as a religion based on austerity, tended to scale down needs and delimit human appetites for material goods. Together they served to slow the pace of technological innovation sufficiently such that when new means of production were developed, they could be sensitively integrated into a balanced society. In medieval times, markets were still modest, usually local affairs, in which guilds exercised strict control over prices, competition, and the quality of the goods produced by their members.

"Grow or Die"

But just as hierarchies and class structures had acquired momentum and permeated much of society, so too the market began to acquire a life of its own and extended its reach beyond a few limited regions into the depths of vast continents. Where exchange had once been primarily a means to provide for essential needs, limited by guilds or by moral and religious restrictions, long-distance trade subverted those limits. Not only did trade place a high premium on techniques for increasing production; it also became the progenitor of new needs, many of them wholly artificial, and gave a tremendous impetus to consumption and the growth of capital. First in northern Italy and the European lowlands, and later—and most decisively— in England during the seventeenth and eighteenth centuries, the production of goods exclusively for sale and profit (the production of the capitalistic commodity) rapidly swept aside all cultural and social barriers to market growth.

By the late-eighteenth and early-nineteenth centuries, the new industrial capitalist class, with its factory system and commitment to limitless expansion, had embarked on its colonization of the entire world, including most aspects of personal life. Unlike the feudal nobility, with its cherished lands and castles, the bourgeoisie had no home but the marketplace and its bank vaults. As a class, it turned more and more of the world into a domain of factories. In the ancient and medieval worlds, entrepreneurs had normally invested profits in land and lived like country gentry, given the prejudices of the times against "ill-gotten" gains from trade. But the industrial capitalists of the modern world spawned a bitterly competitive marketplace that placed a high premium on industrial expansion and the commercial power it conferred, functioning as though growth were an end in itself.

In social ecology it is crucially important to recognize that industrial growth did not and does not result from changes in cultural outlook alone—least of all from the impact of scientific and technological rationality on society. Growth occurs above all from *harshly objective factors* churned up by the expansion of the market itself, *factors that are largely impervious to moral considerations and efforts at ethical persuasion.* Indeed, despite the close association between capitalist development and technological innovation, the most driving imperative of any enterprise in the harshly capitalist marketplace, given the savagely dehumanizing competition that prevails there, is the need of an enterprise to grow in order to avoid perishing at the hands of its savage rivals. Important as even greed may be as a motivating force, sheer survival requires that the entrepreneur must expand his or her productive apparatus in order to remain ahead of others. Each capitalist, in short, must try to devour his or her rivals—or else be devoured by them. The key to this law of life—to survival—is expansion, and the quest for ever-greater profits, to be invested, in turn, in still further expansion. Indeed, the notion of progress, once regarded as faith in the evolution of greater human cooperation and care, is now identified with ever greater competition and reckless economic growth.

The effort by many well-intentioned ecology theorists and their admirers to reduce the ecological crisis to a cultural crisis rather than a social one becomes very obfuscatory and misleading. However ecologically well-meaning an entrepreneur may be, the harsh fact is that his or her very survival in the marketplace precludes the development of a meaningful ecological orientation. The adoption of ecologically sound practices places a morally concerned entrepreneur at a striking and indeed fatal disadvantage in a competitive relationship with a rival—who, operating without ecological guidelines and moral constraints, produces cheap commodities at lower costs and reaps higher profits for further capital expansion. The marketplace has its own law of survival: only the most unscrupulous can rise to the top of that competitive struggle.

Indeed, to the extent that environmental movements and ideologies merely moralize about the wickedness of our anti-ecological society and call for changes in personal lifestyles and attitudes, they obscure the need for concerted social action and tend to deflect the struggle for far-reaching social change. Meanwhile, corporations are skillfully manipulating this popular desire for personal ecologically sound practices by cultivating ecological mirages. So it is that we see, among hundreds of similar advertisements, Mercedes-Benz declaim, in a two-page magazine advertisement, decorated with a bison painting from a Paleolithic cave wall, that "We must work to make progress more environmentally sustainable by including environmental themes in the planning of new products."[7] If such messages are commonplace in Germany, one of western Europe's worst polluters, the same advertising is equally manipulative in the United States, where leading polluters piously declare that for them, "Every Day is Earth Day."

The point social ecology emphasizes is not that moral and spiritual persuasion and renewal are meaningless or unnecessary; they are necessary and can be educational. But modern capitalism is *structurally* amoral and hence impervious to moral appeals. The modern marketplace is driven by imperatives of its own, irrespective of what kind of CEO sits in a corporation's driver's seat or holds on to its handlebars. The direction it follows depends not upon ethical prescriptions and personal inclinations but upon objective laws of profit or loss, growth or death, eat or be eaten, and the like. The maxim "Business is business" explicitly tells us that ethical, religious, psychological, and emotional factors have virtually no place in the predatory world of production, profit, and growth. It is grossly misleading to think that we can divest this harsh, indeed mechanistic world of its objective characteristics by means of ethical appeals.

A society based on the law of "grow or die" as its all-pervasive imperative must of necessity have a devastating impact on first nature. Nor does "growth" here refer to population growth; the current wisdom of population-boomers to the contrary, the most serious disruptors of ecological cycles are found in the large industrial centers of the world, which are not only poisoning water and air but producing the greenhouse gases that are melting the ice caps and threatening to flood vast areas of the planet. Suppose we could somehow cut the world's population in half: would growth and the despoliation of the earth be reduced at all? Capital would insist that it was "indispensable" to own two or three of every appliance, motor vehicle, or electronic gadget, where one would more than suffice if not be too many. In addition, the military would continue to demand ever more lethal instruments of death and devastation, of which new models would be provided annually.

Nor would "softer" technologies, if produced by a grow-or-die market, fail to be used for destructive capitalistic ends. Two centuries ago, large forested areas in England were hacked into fuel for iron forges with axes that had not changed appreciably since the Bronze Age, and ordinary sails guided ships laden with commodities to all parts of the world well into the nineteenth century. Indeed, much of the United States was cleared of its forests, wildlife, and its Indigenous inhabitants with tools and weapons that could have easily been recognized, however much they were modified, by Renaissance people centuries earlier. What modern technics did was *accelerate* a process that had been well under way at the close of the Middle Ages. It cannot be held solely responsible for endeavors that were under way for centuries; it essentially abetted damage caused by the ever-expanding market system, whose roots, in turn, lay in one of history's most fundamental social transformations: the elaboration of a system of production and distribution based on exchange rather than complementarity and mutual aid.

An Ecological Society

Social ecology is an appeal not only for moral regeneration but, and above all, for social reconstruction along ecological lines. It emphasizes that, taken by itself, an ethical appeal to the powers that be, based on blind market forces and ruthless competition, is certain to be futile. Indeed, taken by itself, such an appeal *obscures* the real power relationships that prevail today by making the attainment of an ecological society seem merely a matter of changing individual attitudes, spiritual renewal, or quasi-religious redemption.

Although always mindful of the importance of a new ethical outlook, social ecology seeks to redress the ecological abuses that the prevailing society has inflicted on the natural world by going to the structural as well as the subjective sources of notions like the domination of first nature. That is, it challenges the entire system of

domination itself—its economy, its misuse of technics, its administrative apparatus, its degradations of political life, its destruction
of the city as a center of cultural development, indeed the entire
panoply of its moral hypocrisies and defiling of the human spirit—
and seeks to eliminate the hierarchical and class edifices that have
imposed themselves on humanity and defined the relationship
between nonhuman and human nature. It advances an ethics of complementarity in which human beings play a supportive role in per-
petuating the integrity of the biosphere—the potentiality of human
beings to be the most conscious products of natural evolution.
Indeed, humans have an ethical responsibility to function creatively
in the unfolding of that evolution. Social ecology thus stresses the
need to embody its ethics of complementarity in palpable social
institutions that will make human beings conscious ethical agents in
promoting the well-being of themselves and the nonhuman world.
It seeks the enrichment of the evolutionary process by the diversi-
fication of life-forms and the application of reason to a wondrous
remaking of the planet along ecological lines. Notwithstanding most
romantic views, "Mother Nature" does not necessarily "know best."
To oppose activities of the corporate world does not require one to
become naïvely biocentric. Indeed by the same token, to applaud
humanity's potential for foresight, rationality, and technological
achievement does not make one anthropocentric. The loose usage
of such buzzwords must be brought to a definitive end by reflective
discussion, not by deprecating denunciations.

Social ecology, in effect, recognizes that—like it or not—the
future of life on this planet pivots on the future of society. It con-
tends that evolution, both in first nature and in second, is not yet
complete. Nor are the two realms so separated from each other that
we must choose one or the other—either national evolution, with
its "biocentric" halo, or social evolution, as we have known it up to
now, with its "anthropocentric" halo—as the basis for a creative biosphere. We must go beyond both the natural and the social toward a
new synthesis that contains the best of both. Such a synthesis must
transcend both first and second nature in the form of a creative,
self-conscious, and therefore "free nature," in which human beings

intervene in natural evolution with their best capacities—their ethical sense, their unequaled capacity for conceptual thought, and their remarkable powers and range of communication.

But such a goal remains mere rhetoric unless a *movement* gives it logistical and social tangibility. How are we to organize such a movement? Logistically, "free nature" is unattainable without the decentralization of cities into confederally united communities sensitively tailored to the natural areas in which they are located. Ecotechnologies, solar, wind, methane, and other renewable sources of energy; organic forms of agriculture; and the design of humanly scaled, versatile industrial installations to meet the regional needs of confederated municipalities—all must be brought into the service of an ecologically sound world based on an ethics of complementarity. It means too an emphasis not only on recycling but on the production of high-quality goods that can, in many cases, last for generations. It means the replacement of needlessly insensate labor with creative work and an emphasis on artful craftspersonship in preference to mechanized production. It means the free time to be artful and to fully engage in public affairs. One would hope that the sheer availability of goods, the mechanization of production, and the freedom to choose one's material lifestyle would sooner or later influence people to practice moderation in all aspects of life as a response to the consumerism promoted by the capitalist market.[8]

But no ethics or vision of an ecological society, however inspired, can be meaningful unless it is embodied in a living politics. By *politics*, I do not mean the statecraft practiced by what we call politicians—namely, representatives elected or selected to manage public affairs and formulate policies as guidelines for social life. To social ecology, politics means what it meant in the democratic *polis* of classical Athens some two thousand years ago: direct democracy, the formulation of policies by directly democratic popular assemblies, and the administration of those policies by mandated coordinators who can easily be recalled if they fail to abide by the decision of the assembly's citizens. I am very mindful that Athenian politics, even in its most democratic periods, was marred by the existence of slavery and patriarchy, and by the exclusion of the stranger from public life. In this respect, to be sure, it differed very little from most of the other ancient Mediterranean civilizations—and certainly ancient Asian ones—of the time. What made Athenian politics unique, however, was that it produced institutions that were extraordinarily democratic—even directly so—by comparison with the republican institutions of the so-called "democracies" of today's world. Either directly or indirectly, the Athenian democracy inspired later, more all-encompassing direct democracies, such as many medieval European towns, the little-known Parisian "sections" (or neighborhood assemblies) of 1793 that propelled the French Revolution in a highly radical direction, and more indirectly, New England town meetings and other, more recent attempts at civic self-governance.[9]

Any self-managed community, however, that tries to live in isolation and develop self-sufficiency risks the danger of becoming parochial, even racist. Hence the need to extend the ecological politics of a direct democracy into confederations of ecocommunities, and to foster a healthy interdependence, rather than an introverted, stultifying independence. Social ecology would be obliged to embody its ethics in a politics of libertarian municipalism, in which municipalities conjointly gain rights to self-governance through networks of confederal councils, to which towns and cities would be expected to send their mandated, recallable delegates to adjust differences. All decisions would have to be ratified by a majority of the popular assemblies of the confederated towns and cities. This institutional process could be initiated in the neighborhoods of giant cities as well as in networks of small towns. In fact, the formation of numerous "town halls" has already repeatedly been proposed in cities as large as New York and Paris, only to be defeated by well-organized elitist groups that sought to centralize power rather than allow its decentralization.

Power will always belong to elite and commanding strata if it is not institutionalized in face-to-face democracies, among people who are fully empowered as social beings to make decisions in new communal assemblies. Attempts to empower people in this manner and form constitute an abiding challenge to the nation-state—that is, a dual power in which the free municipality exists in open tension with the nation-state. Power that does not belong to the people invariably belongs to the state and the exploitative interests it represents. Which is not to say that diversity is not a desideratum; to the contrary, it is the source of cultural creativity. Still it never should be celebrated in a nationalistic sense of "apartness" from the general interests of humanity as a whole, or else it will regress into the parochialism of folkdom and tribalism.

Should the full reality of citizenship in all its discursiveness and political vitality begin to wane, its disappearance would mark an unprecedented loss in human development. Citizenship, in the classical sense of the term, which involved a lifelong, ethically oriented education in the art of participation in public affairs (not the empty form of national legitimation that it so often consists of today), would disappear. Its loss would mean the atrophying of a communal life beyond the limits of the family, the waning of a civic sensibility to the point of the shriveled ego, the complete replacement of the public arena with the private world and with private pursuits.

The failure of a rational, socially committed ecology movement would yield a mechanized, aesthetically arid, and administered society, composed of vacuous egos at best and totalitarian automata at worst. Before the planet was rendered physically uninhabitable, there would be few humans who would be able to inhabit it.

Alternatively, a truly ecological society would open the vista of a "free nature" with a sophisticated eco-technology based on solar, wind, and water; carefully treated fossil fuels would be sited to produce power to meet rationally conceived needs. Production would occur entirely for use, not for profit, and the distribution of goods would occur entirely to meet human needs based on norms established by citizens' assemblies and confederations of assemblies. Decisions by the community would be made according to direct, face-to-face procedures with all the coordinative judgments by mandated delegates. These judgments, in turn, would be referred back for discussion, approval, modification, or rejection by the assembly of assemblies (or Commune of communes) *as a whole*, reflecting the wishes of the fully assembled majority.

We cannot tell how much technology will be expanded a few

decades from now, let alone a few generations. The growth and the prospects it is likely to open over the course of this century alone are too dazzling even for the most imaginative utopian to envision. If nothing else, we have been swept into a permanent technological and communications revolution whose culmination it is impossible to foresee. This amassing of power and knowledge opens two radically opposing prospects: either humanity will truly destroy itself and its habitat, or it will create a garden, a fruitful and benign world that not even the most fanciful utopian, Charles Fourier, could have imagined.

It is fitting that such dire alternatives should appear now and in such extreme forms. Unless social ecology—with its naturalistic outlook, its developmental interpretations of natural and social phenomena, its emphasis on discipline with freedom and responsibility with imagination—can be brought to the service of such historic ends, humanity may well prove to be incapable of changing the world. We cannot defer the need to deal with these prospects indefinitely: either a movement will arise that will bestir humanity into action, or the last great chance in history for the complete emancipation of humanity will perish in unrestrained self-destruction.

November, 1984

Market Economy or Moral Economy?

Sooner or later, every movement for basic social change must come to grips with the way people produce the material means of life— their food, shelter, and clothing—and the way these means of life are distributed. To be discreetly reticent about the material sphere of human existence, to loftily dismiss this sphere as "materialistic," is to be grossly insensitive to the preconditions for life itself. Everything we eat to sustain our animal metabolism, every dwelling or garment we use to shelter us from the inclemencies of nature, are normally provided by individuals like ourselves who must work to provision us, as we, one hopes, are obliged to provision them.

Although economists have blanketed this vast activity with amoral, often pretentiously "scientific" categories, preindustrial humanity always saw production and distribution in profoundly moral terms. The cry for "economic justice" is as old as the exis- tence of economic exploitation. Only in recent times has this cry lost its high standing in our notion of ethics or, more precisely, been subordinated to a trivial place by a supraeconomic emphasis on "spirituality" as distinguished from "materiality." Accordingly, it is easy to forgive the great German thinker Theodor Adorno for acidly observing a generation ago: "There is tenderness only in the coarsest demand: that no one shall go hungry anymore."[1]

Overstated as this image of tenderness may seem, it is a much-deserved slap in the faces of those privileged strata whose "chubby insatiability" for the good things of life is matched only by their "chubby insatiability" for the contrived problems of their shriveled and bored egos. It is time—indeed, necessary—to restore the moral dimension of what we so coldly denote as "the economy," and more to the point, to ask what a truly moral economy is.

The difficulty in tying economics to morality stems from the nature of economic life as we know it today. Not that any economy can ever really be "amoral," as the economists or practitioners of "economic science" would have us believe, nor, for that matter, can ways of work and technology ever be regarded as "amoral."[2]

The fact is that our present market economy is grossly *immoral*. Even in denying that economics can be regarded as an authentically moral domain in which people always make decisions about who shall do what, what shall be distributed to whom, and how "scarce resources" shall be weighed against "unlimited needs," the economists have *literally* "demoralized" us and turned us into moral cretins. Price formation, to take only one example, is not merely an impersonal "amoral" computation of supply versus demand. It is an insidious manipulation of both supply and demand—an immoral manipulation of human needs as part of an immoral pursuit of gain. In speaking of a "market economy" as distinguished from a "moral economy," it would not be false to speak of an "immoral economy" as distinguished from a "moral economy."

But this distinction is hard to see, not only because economics, with its panoply of scientistic pretensions, has muddled the entire issue of economics and morality. It is also hard to see because we tend to assume that the economic status quo is a given, a "natural state of affairs," that is assumed to be part of a fictitious "human nature." So deeply rooted is the market economy in our minds that its grubby language has replaced our most hallowed moral and spiritual expressions. We now "invest" in our children, marriages, and personal relationships, a term that is equated with words like "love" and "care." We live in a world of "trade-offs" and we ask for the "bottom line" of any emotional "transaction." We use the terminology of contracts rather than that of loyalties and spiritual affinities. This kind of business babble, garnished with electronic terms like "input," "output," and "feedback," could easily fill a dictionary for our times and those which lie ahead.

Life, in effect, has acquired those descriptive traits that earlier generations once assigned to strictly market interactions—interactions whose influence on their conduct was marginal, however invasive it became in periods of economic difficulty. The "dignity of labor" denoted the subordinate role of work to the higher moral concerns of the worker's sense of self-esteem, however much this dignity was violated by the harshness of toil and the commanding presence of economic hierarchies. "Respect" was a criterion for transactions of any kind, and figured no less in the claims of the workplace militant than it did in those of the Mafia "Godfather." In many countries on the road to industrialization, workers waged strikes to defend their self-esteem and express their moral solidarity, not only to gain material and social benefits.

Today, we have virtually lost this sense of moral direction because our social map has been completely taken over by the market. Our economic coordinates deny us any of the means for comparing ethical images of the past with the gray "amorality" of the present. As recently as the 1930s, people could contrast the "dog-eat-dog" attributes of the marketplace with the solidarity of a village-type neighborhood world and its rich supports in the extended family, whose older members formed living recollections of a more caring preindustrial society. Immediately outside the dense, poisoned cities of the world, the countryside was a visible presence, with traditional agrarian lifeways that were hallowed by the ages. However much one may choose to criticize this archaic refuge from the factory, office, and commercial emporium for its parochialism and patriarchialism, the fact remains that it provided a deeply human and personal refuge—one that was fecund with a limitless capacity for renewal and vitality.

Perhaps equally important, it provided "industrial man" with a sense of contrast and tension between a moral world where values of virtue and the good life guided economic standards, and a marketplace world where values of gain and egotism guided moral standards. This sense of contrast and tension was carried inwardly by workers into shop and home, union and family, factory and neighborhood, city and town. Even when the market economy seemed to be the focal center of life during the working day, a sense of an older, more congenial and moral world to which one could later repair existed in the peripheral vision of the ordinary worker. The space to be a human being with spontaneous human concerns clashed with the space that forced the individual to be a class being, a creature of the market economy and its highly rationalized industrial core.

Ironically, in the vision of millions, the Great Depression of the 1930s moved the market economy from its primary status in the previous decade to a secondary one. Despite the prevalence of a naïve commitment to progress and belief in the power of technology to remove all the ills of society, the generation of the early thirties moved in great numbers from the city to the countryside, tightened its family bonds to meet economic adversity, intensified its sense of local solidarity and, with it, neighborhood and town support systems. In short, it recovered moral commitments between people despite the great dislocations that occurred among farmers in the drought-stricken prairies of North America's Dust Bowl and the torrential increase in the urban displaced who filled the railroad box cars of the middle and far West.

As a result of this parallel movement into and out of the centers of industry and commerce, the impersonal world of frenzied speculation and paper riches so exuberantly celebrated during the boom years of the 1920s suffered a major loss of prestige, as the revival of populist and socialist movements so clearly revealed. The stock market collapse in 1929 ended a popular reverence not only for corporate wealth, but also for the market system itself. Barter, mutual aid, the verities of an agrarian America, self-reliance, and independence, together with regionalism and cultural identity, haunted the land for years and even invaded its artistic canons, as witness the paintings of Grant Wood, the WPA muralists and photographers, and the resurgence of research into local lore and traditions.

Today, this decade-long lapse of the market economy's prestige has simply been forgotten. From the 1950s onward, the market economy has not only imperialized every aspect of conventional life, it has also dissolved the memory of the alternative lifeways that precede it. We are all anonymous buyers and sellers these days, even of the miseries that afflict us. We not only buy and sell our labor power in all its subtle forms, we buy and sell our neuroses, anomie, loneliness, spiritual emptiness, integrity, lack of self-worth, and emotions, such as they are, to gurus, specialists in mental and physical "well-being," psychoanalysts, clerics in all garbs, and ultimately to the armies of corporate and governmental bureaucrats who have finally become the authentic sinews of what we euphemistically call "society." We buy and sell the outward trappings of personality: the sheen-like leather jackets that make humble bookkeepers look like dashing pimps and the high-heeled boots that make bored secretaries look like dangerously seductive temptresses. Clothing, face paint, well-blown coiffures, baubles, a vast array of insignia and tokens all combine in the urban cesspools of the world to make us seem more "interesting" and less depersonalized than we really are.

Convention submerges in a quick dip only to resurface as stylized indiosyncracies, damning badges of "individuation" that subtly affirm its loss. The snapped cap of the traditional worker, even the high hat of the cartoon bourgeois, once topped faces that were etched with character, experience, inner strength, and individuality. Today the doll-like heads of our "bohemian" middle classes, these relics of a vibrant past, seem like grotesque caricatures. Today the market economy has shown its power to reach the most inward recesses of personality by making its acolytes into look-alikes even as they grasp for the idiosyncratic in dress and the low culture of the mass media. Indeed, whatever is culturally exciting and fills our concert halls and theaters to the bursting point is the recycled product of generations now dead or dying—often recycled with a technical proficiency and slickness that bleeds it of all character and earthiness.

Our liberalism toward every moral excess seems more like indifference than tolerance. Anomic, spiritless, and unfeeling, we have become the very free-floating commodities we so eagerly produce and devour. Society, in turn, flattened and colorless, has become the very market economy we once confined to the personally remote world of "business." The immorality of our credo of "amorality" stems from a sense of indifference that is evil because it has no criteria for the good and the virtuous. Its philosophy consists of the endless prattle of small talk and its ideals are embodied in its garishly cluttered shopping malls, which have become its most imperious and sacred temples.

The market economy is blessed with a grand secret from which it draws its power to shape the totality of social life: the power of anonymity. Sellers do not know buyers and buyers do not know sellers. What sellers dump on the market—all self-serving myths of "salesmanship" aside—are their commodities, not themselves. A buyer who purchases a dress ultimately confronts an object, a dress—not its producer, a person. Admittedly, there are producers who fit a buyer for a garment and "sales" personnel who oil the purchase along. But the fitter or tailor is a marketplace archaism who actually belongs to a bygone era, or serves a highly affluent elite. The "salesperson" is at best a catalyst for making purchasable dreams more palatable. He or she is virtually nonexistent in the great shopping malls, where the principal encounter between buyer and seller occurs on a checkout line at a cash register, not in the more intimate world where the purveyor of merchandise tries to persuade a potential buyer into a purchase. No, the market economy is structured around buyer and object, or producer and retail establishment, not between person and person.

The anonymity of the exchange process today has formidable consequences, more far-reaching than we normally suspect. We are struck first by its suffocating impersonality. A machine called the market takes over vital functions that rightly should be performed by the intercourse between people. Although electronic and print media continually barrage us with images and voices that seem like human beings, we rarely encounter real flesh-and-blood people in the modern market. Often, no way exists to leisurely discuss the worthiness of a product with the producer who, it would seem, can best judge its qualities and utility. Salespersons, few as they are, are notoriously ignorant about the commodities they purvey and can be easily outwitted by any knowledgeable buyer. Moreover, they are generally outrageously indifferent and excessively rehearsed. They can be—in some places have already been—replaced by a recording. In the impersonality of the market, no interchange between buyer and seller exists that can lend itself to *ethical* guidance.

In all past eras, the worthiness of a product was morally integrated with the worthiness of its seller and producer. The value that a buyer placed on a commodity, indeed, on any exchangeable entity, constituted an ethical gauge of the moral integrity of the individual from whom it was acquired. To denigrate this object, to return it with disparaging remarks about its quality, was to impugn the seller's probity and self-esteem—not simply as a "good" producer, but as a person with ethical standards. The craftsperson, in this sense, was as "good" as the "goods" he or she crafted; the seller, as "good" as "goods" he or she sold. I use the word "good" not instrumentally, in terms of technical proficiency—a word that today, quite characteristically, usually means precisely that—but ethically, in terms of human "goodness" and moral probity. "Good will" meant honesty, integrity, reliability, responsibility, and a high sense of public service, rather than staying power in the marketplace jungle, fiscal soundness, and the contrived myth of "superiority" inculcated in the public mind by advertising. One did not buy a "name" that repeatedly appeared on television screens, neon signs, and billboards; one "bought" the moral certainty of a good personal reputation, an artist's sense of commitment to aesthetic excellence, the cherished *aretē*, or virtue, that the Greeks imputed to an individual's vocation as a moral calling, and the deeply felt responsibility of a good worker to a product that constituted an extension of his or her human powers. "Goods" and "goodness"—a commonality of terms that is not accidental—carried the ethical imprimatur of social responsibility, not the instrumental slickness of technical finesse and hard-sell.

The actual act of selling, in turn, had its own etiquette and personal ambience. Buyer and seller encountered each other with talk about the affairs of the day, personal inquiries and assurances, opinions on a host of public issues, and finally, a mutual interest in the product, with knowledgeable remarks about its components, artwork, and merits. A price was a moral bond, not a mere exchange of "goods" for money. The signature of the producer or seller appeared on the product as well as the bill of sale. People used terms like "just prices," not simply "bargains." Between buyer and seller was an ethical tie that signified their reliance, indeed their *dependence*, on each other for the needful and good things of life. A high sense of mutuality, based on trust and a shared recognition of faith in a nexus of complementarity for sustaining survival itself, permeated the entire exchange process.

We should not consign such relationships to distant ages like the medieval world. However vestigial in form, they existed as recently as the 1930s, when production, despite its increasingly mass character, was commonly tested in the deeply personal arena of small neighborhood retail shops; in the fitting rooms of garment makers; in cobbler, cigar-making, and bakery shops; and in an endless array of service establishments where work was done under the eyes of the customer and even under the eyes of passing crowds.

Today, the anonymity and depersonalization of the market has almost completely divested the exchange process of this moral dimension. Even in so-called alternative enterprises like organic farms, craft shops, and food cooperatives, the ethical inspiration which presumably gave rise to them has been gravely diluted and threatens to fade away. To the degree that these establishments become "established," they become more entrepreneurial than moral. This is especially true when moral inspiration is confused with material need. An agribusiness organic farm that is meant merely to satisfy a "need" for "good food" rather than food that is cultivated from a sense of "goodness" and ecological concern—like a "food cooperative" that is meant to provide "good food" at cheap prices—is guided more by need than by ethics. That is to say, it is meant to satisfy a concern that is pragmatic rather than moral.

Ironically, none of these concerns can ever supplant the shopping mall. No local organic farm can compete successfully with agribusiness, and no food cooperative can successfully outbid, much less outsupply, a supermarket. The most these "alternative" enterprises can do is to coexist precariously with the giants that tower over them, as mere marginalia that appeal on strictly material grounds to society's fringes, not society at large.

Worse, as practical projects that aim for "efficiency," "high returns," expanded operations, and a more "successful" marketing strategy, they may begin to objectify their consumers as much as they do the produce they sell. They become merely another impersonal business enterprise whose "goods" are as lacking in "goodness" as those of their larger rivals. Dwarfed by the giants who smirk at their existence and claims, they become food pharmacies for dispensing unpolluted "organic" products instead of pills—the drugs for coping with a social disease, not for preventing or curing it. In short, they can become as inorganic, depersonalized, computerized, and cynical as the larger enterprises on whose turf they nibble—dumping grounds for organic foods to meet the therapeutic needs of an increasingly anonymous and inorganic public. The moral aspects of distributing or growing food and other produce are blotted out by considerations of "efficiency" and "success"—the two attributes of capitalistic enterprise that lend themselves to a concern for economic quantity at the expense of ethical quality.

To put the issue bluntly: an organic carrot, a homespun garment, a crafted plank of wood, or a hand-worked leather boot is merely a "thing" that people confront as impersonally in a food cooperative or a craft shop as they do in a shopping mall *if it does not carry a moral message that changes it as an exotic creature of an immoral economy.* The "thing" itself will never give voice to a moral message merely by its quality, ecological pedigree, and usefulness. As wholesome, nourishing, attractive, and free of the pollutants that infect our bodies and tastes as it may be, it does not become a "good" in a moral sense for these reasons alone. Moral "goodness" can come only from the *way* in which people interact between themselves, and the sense of ethical purpose they give to their productive activities. It is through the way "goods" are exchanged or, to state the case more radically, the way exchange is used to appropriately distribute them such that "buyer" and "seller" cease to be polarized against each other and are joined in an economic community, united by a fraternal or sororal relationship based on a sense of mutual identification and personal complementarity. Care, responsibility, and obligation become the authentic "price tag" of the moral economy, as distinguished from the interest, cost, and profitability that enter into the "price tag" of the market economy.

Care, responsibility, and obligation, we are told, are "ideological" concepts which have no place in a scientistic notion of economics. This criticism points to the very heart of the issues raised by a moral economy. A moral economy—a participatory system of distribution based on ethical concerns—is meant to dissolve the immorality that the modern mind identifies with economics as such. Its goal is to dissolve the antagonism between "buyer" and "seller," to show that in practice both "buyer" and "seller" form a *community* based on a rich sense of mutuality, not on the opposition of "scarce resources" to "unlimited needs." The object exchanged is secondary to the ethical values that are explicitly shared by the participants of a moral economy. For "buyer" and "seller" to care for each other's well-being, for them to feel deeply responsible to each other, and for them to be cemented by a deep sense of obligation for their mutual welfare is to replace a strictly economic nexus with an ethical one—that is, *to turn economics into culture* rather than to visualize it as the "circulation" of things. Where distribution becomes a form of complementarity, it ceases in fact to be economic in the usual meaning of the word and the terms "buyer" and "seller" become meaningless.

Material needs begin to express one of many ways in which claims for things become claims for moral integrity. The "buyer's" expectations begin to expand beyond mere needs to a belief in the "seller's" ability to exhibit the highest moral probity in providing the material means of life. The "seller," in turn, advances his or her goods, and "goodness"—an ethical conviction that the means of life serve to satisfy not only material needs, but also spiritual ones that foster trust, community, and solidarity. The rivalry and seeming independence that pervades the market economy is replaced by reciprocity and interdependence in which distribution with its moral etiquette—like primitive rituals—affirms a sense of unity and shared destiny between its participants. The inequalities conferred by differences in strength, health, age, and skill cease to be the damning stigmas of a specious "equality" that permits each individual to drift on his or her own in a deadening and emotionally blunted pursuit of advantage. To the contrary, they spawn a sense of complementarity and a commitment to compensation that yields the great radical maxim: From each according to his or her abilities, to each according to his or her needs.

These images of a moral economy and its ethical preconditions are not abstractions. They imply concrete institutions and specific forms of behavior. Institutionally, they presuppose a new form of productive community, as distinguished from a mere marketplace where each buyer and seller fends for herself or himself—a community in which actual producers are networked and interlocked somewhat like the old medieval guilds in a responsible support system. In this support system, the producers—be they organic farmers, carpenters, leather workers, jewelers, weavers, clothiers, builders, craftspeople and shop workers of all kinds, indeed, professionals such as physicians, chiropractors, nurses, attorneys, teachers, and the like—explicitly agree to exchange their products and services on terms that are not merely "equitable" or "fair" but supportive of each other. Like all real communities, they form a family that provides for the welfare of its participants as a collective responsibility, not simply a personal responsibility. For example, medical people assume a moral duty to care for the health needs of craftspeople, who in turn assume the task of provisioning the community's physicians, nurses, dieticians, etc. This sense of moral complementarity—this social "ecosystem," so to speak—encompasses all members of the productive community. Price, resources, personal interests, and costs play no role in a moral economy. Services and provisions are available as needed, with no "accounting" of what is given and taken.

"Need," in turn, is moralized in the very profound sense of a shared concern of the giver as well as the receiver, for it becomes important for the producer of a "good" to see to it that the consumer suffers no privation or want for lack of his or her product, indeed, that the "good" is the *best* that can be given to whoever is needful. To go "beyond good and evil," if I may use the title of Nietzsche's provocative work, is to seek excellence for its own sake and, above all, for the community's sake rather than remain trapped in amorality or moral relativism.

"Need" turns from mere want of a "good" into a way of identifying producer and consumer in a caring social bond that is guided not by interest, profitability, and cost, with all their quantitative trappings, but by that ineffable qualitative and disinterested sense of mutual welfare such as we expect in parental and sibling relationships. It is no longer the yearning of one individual for a "good," but a collective funding of desire with the shared expectation that fulfillment is a communal desideratum, just as a lover experiences the joy of the beloved in the very fact that a desire is satisfied. Inasmuch as virtually every consumer is in some sense a producer, the fictive opposition between consumption and production, with its connotation of the "innocent consumer" who must be protected from the "predatory producer," is eliminated.

That the infirm, elderly, or very young do not seem to belong to such a productive community in the technical sense is perhaps all the more reason to include them fully in its benefits, if only to test continually the moral intentions of such a community—that is, to confront it with an ongoing challenge of its own moral integrity and disinterestedness. And yet even the elderly and the infirm, I suspect, will *want* to find a function for themselves in a moral economy, be it simply custodial, clerical, or instructive, depending on their training and background in the more active periods of their lives. The point is that a moral economy exists for moral reasons, not simply for reasons of survival or gain. The good life, materially supported by "goods" that are the messengers of "goodness," is an end in itself: a source of new selfhood and new ways of life; an ongoing education in forms of association, virtue, and decency; a countervailing force to the socially, morally, and psychologically corrosive marketplace and its unbridled egotism.

Such a moral economy has no historical precedents on which to model itself—and, in a very real sense, can only be created by practice and experience, rather than precept and past example. But its architects can draw some inspiration from many Indigenous communities in which usufruct, not ownership, guided people in the availability of tools and resources.³ Possibly, too, they can learn from the democratic guild forms of organization that existed in early medieval townships and from certain cooperative or quasi-religious forms of productive association like the Hutterite and Tolstoyan communes. But these forms of associations are hints, often defective when taken by themselves and useful when selectively pieced together, of what must ultimately be a broader concept of a moral economy for society as a whole. A moral economy, structurally speaking, may for a long time be a marginal example of what the human community as a whole should one day become. But so much that now exists in the center of human affairs formerly developed on their margins that we should not despair that a moral economy can only be peripheral to society today.

Even more fundamental than structure is the problem of behavior. A moral economy, based on shared concern rather than private interest, is no better than the sensibilities it fosters. If our concept of a material *good* comes from a waning sense of moral *goodness*, the recovery of the tie between the material and the moral, between *good* and *goodness*, recasts our very notion of an economy in a radically new light. It places upon a moral economy the crucial function of developing an economic community into an arena for ethical education, as well as a moral system of production and distribution.⁴

Like the Athenian *polis* of some two thousand years ago, a moral economy must become a school for creating a new kind of citizenship: economic citizenship as well as political, productive citizenship as well as participatory, a place for learning a respect for "things" as products of a fecund nature as well as a center for dedicated work, and the embodiment of a spiritualized physicality as well as a productive domain for creating objects for personal consumption. The "curriculum" for such a school involves a "respiritization" of the work process, the "raw materials" this process shapes, the moral context in which people work together, and the purposes for which they work—this, aside from the more obvious issues of familial, communal, or distinctly pedagogical institutions and politically libertarian forms of self-governance through which people are educated. Hence, the economic arena becomes a "school"—as it has always been, more for the worse than for the better—forming the moral character of the individual as well as providing major guidelines for his or her behavior.

This economic image of moral self-development is inseparable from the tools and machines that give it reality. Ecotechnologies, such as small-scale solar and wind-power devices, ecological agriculture, aquacultural techniques, energy-conserving shelters and devices, in short, that entire panoply of so-called appropriate technologies (a term I find difficult to accept because the word "appropriate"—for what?—is too morally ambiguous) should be seen more in terms of their ethical function than their operational efficiency. That we must bring the sun, wind, land, flora, fauna, and the building materials of our shelters into our lives in a new, eco logically oriented way if we are to develop an authentic respect for the natural world, its fecundity, and our dependence on it should be obvious. There is more to ecotechnology than its efficiency and renewability: our metabolism with nature will either be mutually interdependent such that our vision of ourselves will place us firmly within the natural world—not "above" it—or we will become its most destructive parasites.

Fundamental to that sense of interdependence is a re-visioning of nature as a moral basis for a new ecological ethics. This moral basis, so suspect to the modern scientistic mind, forms the stuff of social ecology and requires separate discussion. Here it suffices to point out that we will either re-envision nature as a domain of fecundity and development or, in the marketplace mentality, conceive of it as a rank jungle to be savagely exploited as we exploit each other in the buyer-seller relationship. A market economy and a moral economy thus stand counterposed to each other on many different levels: in their images of nature, technology, education, work, the production and distribution of the means of life, community, and "goods" as commodities or the embodiment of "goodness."

Above all, they stand counterposed to each other in the way men and women envision themselves and the ideals they advance for human intercourse—indeed, whether these ideals advance no further than mere survival, with all its narrow technocratic and ethical implications, or rise to the level of life, with its broad ecological and ethical implications. On this score, a market economy and a moral economy raise fundamentally opposed notions of humanity's self-realization and sense of purpose, concepts which define the very meaning of material premises on which our development eventually depends.

July, 1983

Radical Politics in an Era of Advanced Capitalism

Defying all the theoretical predictions of the 1930s, capitalism has restabilized itself with a vengeance and acquired extraordinary flexibility in the decades since World War II. In fact, we have yet to clearly determine what constitutes capitalism in its most "mature" form, not to speak of its social trajectory in the years to come. But what is clear, I would argue, is that capitalism has transformed itself from an *economy* surrounded by many precapitalist social and political formations into a *society* that itself has become "economized."

Terms like *consumerism* and *industrialism* are merely obscurantist euphemisms for an all-pervasive embourgeoisement that involves not simply an appetite for commodities and sophisticated technologies but the expansion of commodity relationships—of market relationships—into areas of life and social movements that once offered some degree of resistance to, if not a refuge from, utterly amoral, accumulative, and competitive forms of human interaction. Marketplace values have increasingly percolated into familial, educational, personal, and even spiritual relationships and have largely edged out the precapitalist traditions that made for mutual aid, idealism, and moral responsibility in contrast to businesslike norms of behavior.

There is a sense in which any new forms of resistance—be they by left libertarians, or radicals generally—must open alternative areas of life that can countervail and undo the embourgeoisement of society at all its levels. The issue of the relationship of "society," "politics," and "the state" becomes one of programmatic urgency. Can there be any room for a radical public realm beyond the communes, cooperatives, and neighborhood service organizations fostered by the 1960s counterculture—structures that easily degenerated into boutique-type businesses when they did not disappear completely? Is there, perhaps, a public realm that can become an arena for the interplay of conflicting forces for change, education, empowerment, and ultimately, confrontation with the established way of life?

Marxism, Capitalism, and the Public Sphere

The very concept of a *public realm* stands at odds with traditional radical notions of a *class realm*. Marxism, in particular, denied the existence of a definable "public," or what in the Age of Democratic Revolutions of two centuries ago was called "the People," because the notion ostensibly obscured specific class interests—interests that were ultimately supposed to bring the bourgeoisie into unrelenting conflict with the proletariat. If "the People" meant anything, according to Marxist theorists, it seemed to mean a waning, unformed, nondescript petty bourgeoisie—a legacy of the past and of past revolutions—that could be expected to side mainly with the capitalist class it aspired to enter and ultimately with the working class it was forced to enter. The proletariat, to the degree that it became class conscious, would ultimately express the general interests of humanity once it absorbed this vague middle class, particularly during a general economic or "chronic" crisis within capitalism itself.

The 1930s, with its waves of strikes, its workers' insurrections, its street confrontations between revolutionary and fascist groups, and its prospect of war and bloody social upheaval, seemed to confirm this vision. But we cannot any longer ignore the fact that this traditional radical vision has since been replaced by the present-day reality of a managed capitalist system—managed culturally and ideologically as well as economically. However much living standards have been eroded for millions of people, the unprecedented fact remains that capitalism has been free of a "chronic crisis" for a half-century. Nor are there any signs that we are faced in the foreseeable future with a crisis comparable to that of the Great Depression. Far from having an internal source of long-term economic breakdown that will presumably create a general interest for a new society, capitalism has been more successful in crisis management in the last fifty years than it was in the previous century and a half, the period of its so-called "historical ascendancy."

The classical industrial proletariat, too, has waned in numbers in the First World (the historical *locus classicus* of socialist confrontation with capitalism), in class consciousness, and even in political consciousness of itself as a historically unique class. Attempts to rewrite Marxian theory to include salaried people in the proletariat are not only nonsensical, they stand flatly at odds with how this vastly differentiated middle-class population conceives itself and its relationship to a market society. To live with the hope that capitalism will "immanently" collapse from within as a result of its own contradictory self-development is illusory as things stand today.

But there are dramatic signs that capitalism, as I have emphasized elsewhere, is producing external conditions for an ecological crisis—that may well generate a general human interest for radical social change. Capitalism, organized around a "grow-or-die" market system based on rivalry and expansion, must tear down the natural world—turning soil into sand, polluting the atmosphere, changing the entire climatic pattern of the planet, and possibly making the earth unsuitable for complex forms of life. In effect, it is proving to be an ecological cancer and may well simplify complex ecosystems that have been in the making for countless aeons.

If mindless and unceasing growth as an end in itself—forced by competition to accumulate and devour the organic world—creates problems that cut across material, ethnic, and cultural differences, the concept of "the People" and of a "public sphere" may become a living reality in history. Some kind of radical ecology movement has yet to be established that could acquire a unique, cohering, and political significance to replace the influence of the traditional workers' movement. If the *locus* of proletarian radicalism was the factory, the *locus* of the ecology movement would be the community: the neighborhood, the town, and the municipality. A new alternative, a political one, would have to be developed that is neither parliamentary on the one hand nor locked into direct action and countercultural activities on the other. Indeed, direct action could mesh with this new politics in the form of community assemblies oriented toward a fully participatory democracy—in the highest form of direct action, the full empowerment of the people in determining the destiny of society.

Society, Politics, and the State

If the 1960s gave rise to a counterculture to resist the prevailing culture, the following decades have created the need for popular counter-institutions to countervail the centralized state. Although the specific form that such institutions could take may vary according to the traditions, values, concerns, and culture of a given area, certain basic theoretical premises must be clarified if one is to advance the need for new institutions and, more broadly, for a new radical *politics*. The need once again to define politics—indeed, to give it a broader meaning than it has had in the past—becomes a practical imperative. The ability and willingness of radicals to meet this need may well determine the future of social movements and the very possibility of radicalism to exist as a coherent force for basic social change.

The major institutional arenas—the social, the political, and the statist—were once clearly distinguishable from each other. The social arena could be clearly demarcated from the political, and the political, in turn, from the state. But in our present, historically clouded world, these have been blurred and mystified. Politics has been absorbed by the state, just as society has increasingly been absorbed by the economy today. If new, truly radical movements to deal with ecological breakdown are to emerge and if an ecologically oriented society is to end attempts to dominate nature as well as people, this process must be arrested and reversed.

It is easy to think of society, politics, and the state ahistorically, as if they had always existed as we find them today. But the fact is that each one of these has had a complex development, one that should be understood if we are to gain a clear sense of their importance in social theory and practice. Much of what we today call *politics*, for one, is really statecraft, structured around staffing the state apparatus with parliamentarians, judges, bureaucrats, police, the military, and the like, a phenomenon often replicated from the summits of the state to the smallest of communities. But the term *politics*, Greek etymologically, once referred to a public arena peopled by conscious citizens who felt competent to directly manage their own communities, or *poleis*.

Society, in turn, was the relatively private arena, the realm of familial obligation, friendship, personal self-maintenance, production, and reproduction. From its first emergence as merely human group existence to its highly institutionalized forms, which we properly call society, social life was structured around the family or *oikos*. (Economy, in fact, once meant little more than the management of the family.) Its core was the domestic world of woman, complemented by the civil world of man.

In early human communities, the most important functions for survival, care, and maintenance occurred in the domestic arena, to which the civil arena, such as it was, largely existed in service. A tribe (to use this term in a very broad sense to include bands and clans) was a truly social entity, knitted together by blood, marital, and functional ties based on age and work. These strong centripetal forces, rooted in the biological facts of life, held these eminently social communities together. They gave them a sense of internal solidarity so strong that the tribes largely excluded the "stranger" or "outsider," whose acceptability usually depended upon canons of hospitality and the need for new members to replenish warriors when warfare became increasingly important.

A great part of recorded history is an account of the growth of the male civil arena at the expense of this domestic or social one.

Males gained growing authority over the early community as a result of intertribal warfare and clashes over territory in which to hunt. Perhaps more important, agricultural peoples appropriated large areas of the land that hunting peoples required to sustain themselves and their lifeways.

It was from this undifferentiated civil arena (again, to use the word *civil* in a very broad sense) that politics and the state emerged. Which is not to say that politics and statecraft were the same from the beginning. Despite their common origins in the early civil arena, these two were sharply opposed to each other. History's garments are never neat and unwrinkled. The evolution of society from small domestic social groups into highly differentiated, hierarchical, and class systems whose authority encompassed vast territorial empires is nothing if not complex and irregular.

The domestic and familial arena itself—that is to say, the social arena—helped to shape the formation of these states. Early despotic kingdoms, such as those of Egypt and Persia, were seen not as clearly civil entities but as the personal "households" or domestic domains of monarchs. These vast palatial estates of "divine" kings and their families were later carved up by lesser families into manorial or feudal estates. The social values of present-day aristocracies are redolent of a time when kinship and lineage, not citizenship or wealth, determined one's status and power.

The Rise of the Public Domain

It was the Bronze Age "urban revolution," to use V. Gordon Childe's expression, that slowly eliminated the trappings of the social or domestic arena from the state and created a new terrain for the political arena. The rise of cities—largely around temples, military fortresses, administrative centers, and interregional markets—created the basis for a new, more secular and more universalistic form of political space. Given time and development, this space slowly evolved an unprecedented public domain.

Cities that are perfect models of such a public space do not exist in either history or social theory. But some cities were neither pre-dominantly social (in the domestic sense) nor statist, but gave rise to an entirely new societal dispensation. The most remarkable of these were the seaports of ancient Hellas and the craft and commercial cities of medieval Italy, Russia, and central Europe. Even modern cities of newly forming nation-states like Spain, England, and France developed identities of their own and relatively popular forms of citizen participation. Their parochial, even patriarchal attributes should not be permitted to overshadow their universal humanistic attributes. From the Olympian standpoint of modernity, it would be as petty as it would be ahistorical to highlight failings that cities shared with nearly all "civilizations" over thousands of years.

What should stand out as a matter of vital importance is that these cities created the public domain. There, in the agora of the Greek democracies, the forum of the Roman republic, the town center of the medieval commune, and the plaza of the Renaissance city, citizens could congregate. To one degree or another in this public domain a radically new arena—a political one—emerged, based on limited but often participatory forms of democracy and a new concept of civic personhood, the citizen.

Defined in terms of its etymological roots, *politics* means the management of the community or *polis* by its members, the citizens. *Politics* also meant the recognition of civic rights for strangers or "outsiders" who were not linked to the population by blood ties. That is, it meant the idea of a universal *humanitas*, as distinguished from the genealogically related "folk." Together with these fundamental developments, politics was marked by the increasing secularization of societal affairs, a new respect for the individual, and a growing regard for rational canons of behavior over the unthinking imperatives of custom.

I do not wish to suggest that privilege, inequality of rights, super-natural vagaries, custom, or even mistrust of the "stranger" totally disappeared with the rise of cities and politics. During the most radical and democratic periods of the French Revolution, for example, Paris was rife with fears of "foreign conspiracies" and a xenophobic mistrust of "outsiders." Nor did women ever fully share the freedoms enjoyed by men. My point, however, is that something very new was created by the city that cannot be buried in the folds of the social or of the state: namely, a public domain. This domain narrowed and expanded with time, but it never completely disappeared from history. It stood very much at odds with the state, which tried in varying degrees to professionalize and centralize power, often becoming an end in itself, such as the state power that emerged in Ptolemaic Egypt, the absolute monarchies of seventeenth-century Europe, and the totalitarian systems of rule established in Russia and in China in the past century.

The Importance of the Municipality and the Confederation

The abiding physical arena of politics has almost always been the city or town—more generically, the municipality. The size of a politically viable city is not unimportant, to be sure. To the Greeks, notably Aristotle, a city or *polis* should not be so large that it cannot deal with its affairs on a face-to-face basis or eliminate a certain degree of familiarity among its citizens. These standards, by no means fixed or inviolable, were meant to foster urban development along lines that directly countervailed the emerging state. Given a modest but by no means small size, the *polis* could be arranged institutionally so that it could conduct its affairs by rounded, publicly engaged men with a minimal, carefully guarded degree of representation.

To be a political person, it was supposed, required certain material preconditions. A modicum of free time was needed to participate in political affairs, leisure that was probably supplied by slave labor, although it is by no means true that all active Greek citizens were slaveowners. Even more important than leisure time was the need for personal training or character formation—the Greek notion of *paidaeia*—which inculcated the reasoned restraint by which citizens maintained the decorum needed to keep an assembly of the people viable. An ideal of public service was necessary to outweigh narrow, egoistic impulses and to develop the ideal of a general interest. This was achieved by establishing a complex network of relationships, ranging from loyal friendships—the Greek notion of *philia*—to shared experiences in civic festivals and military service.

But politics in this sense was not a strictly Hellenic phenomenon. Similar problems and needs arose and were solved in a variety of ways in the free cities not only in the Mediterranean basin but in continental Europe, England, and North America. Nearly all these free cities created a public domain and a politics that were democratic to varying degrees over long periods of time. Deeply hostile to centralized states, free cities and their federations formed some of history's crucial turning points in which humanity was faced with the possibility of establishing societies based on municipal confederations or on nation-states.

The state, too, had a historical development and cannot be reduced to a simplistic ahistorical image. Ancient states were historically followed by quasi-states, monarchical states, feudal states, and republican states. The totalitarian states of this century beggar the harshest tyrannies of the past. But essential to the rise of the nation-state was the ability of centralized states to weaken the vitality of urban, town, and village structures and replace their functions by bureaucracies, police, and military forces. A subtle interplay between the municipality and the state, often exploding in open conflict, has occurred throughout history and has shaped the societal landscape of the present day. Unfortunately, not enough attention has been given to the fact that the capacity of states to exercise the full measure of their power has often been limited by the municipal obstacles they encountered.

Nationalism, like statism, has so deeply imprinted itself on modern thinking that the very idea of a municipalist politics as an option for societal organization has virtually been written off. For one thing, as I have already emphasized, politics these days has been identified completely with statecraft, the professionalization of power. That the political realm and the state have often been in sharp conflict with each other—indeed, in conflicts that exploded in bloody civil wars—has been almost completely overlooked. The great revolutionary movements of the past, from the English Revolution of the 1640s to those in our own century, have always been marked by strong community upsurges and depended for their success on strong community ties. That fears of municipal autonomy still haunt the nation-state can be seen in the endless arguments that are brought against it. Phenomena as "dead" as the free community and participatory democracy should presumably arouse far fewer opponents than we continue to encounter.

The rise of the great megalopolis has not ended the historic quest for community and civic politics, any more than the rise of multinational corporations has removed the issue of nationalism from the modern agenda. Cities like New York, London, Frankfurt, Milan, and Madrid can be *politically* decentralized institutionally, be they by neighborhood or district networks, despite their large structural size and their internal interdependence. Indeed, how well they can function if they do not decentralize structurally is an ecological issue of paramount importance, as problems of air pollution, adequate water supply, crime, the quality of life, and transportation suggest.

History has shown very dramatically that major cities of Europe with populations approaching a million and with primitive means of communication functioned by means of well-coordinated decentralized institutions of extraordinary political vitality. From the Castilian cities that exploded in the *Comuñero* revolt in the early 1500s through the Parisian sections or assemblies of the early 1790s to the Madrid Citizens' Movement of the 1960s (to cite only a few), municipal movements in large cities have posed crucial issues of where power should be centered and how societal life should be managed institutionally.

That a municipality can be as parochial as a tribe is fairly obvious—and is no less true today than it has been in the past. Hence, any municipal movement that is not confederal—that is to say, that does not enter into a network of mutual obligations to towns and cities in its own region—can no more be regarded as a truly political entity in any traditional sense than a neighborhood that does not work with other neighborhoods in the city in which it is located. Confederation—based on shared responsibilities, full accountability of confederal delegates to their communities, the right to recall, and firmly mandated representatives—forms an indispensable part of a new politics. To demand that existing towns and cities replicate the nation-state on a local level is to surrender any commitment to social change as such.

What is of immense practical importance is that prestatist institutions, traditions, and sentiments remain alive in varying degrees throughout most of the world. Resistance to the encroachment of oppressive states has been nourished by village, neighborhood, and town community networks, as witness such struggles in South Africa, the Middle East, and Latin America. To ignore the communal basis of this resistance would be as myopic as to ignore the latent instability of every nation-state. Worse would be to take the nation-state for granted as it is, and deal with it merely on its own terms. Indeed, whether a state remains "more" of a state or "less"—no trifling matter to radical theorists as disparate as Bakunin and Marx—depends heavily upon the power of local, confederal, and community movements to countervail it and hopefully establish a dual power that will replace it. The major role that the Madrid Citizens' Movement played nearly three decades ago in weakening the Franco regime would require a major study to do it justice.

Notwithstanding Marxist visions of a largely economistic conflict between "wage labor and capital," the revolutionary working-class movements of the past were not simply industrial movements. The volatile Parisian labor movement, largely artisanal in character, for example, was also a *community* movement that was centered on quartiers and nourished by a rich neighborhood life. From the Levellers of seventeenth-century London to the anarcho-syndicalists of Barcelona in the twentieth century, radical activity has been sustained by strong community bonds, a public sphere provided by streets, squares, and cafes.

The Need for a New Politics

This municipal life cannot be ignored in radical practice and must even be recreated where it has been undermined by the modern state. A new politics, rooted in towns, neighborhoods, cities, and regions, forms the only viable alternative to the anemic public participation allowed under a system of representative party politics. The duration of strictly single-issue movements, too, is limited to the problems they are opposing. Militant action around such issues should not be confused with the long-range radicalism that is needed to change consciousness and ultimately society itself. Such movements flare up and pass away, even when they are successful. Unless they are embedded in a radical municipalist approach, they lack the institutional underpinnings that are so necessary to create lasting movements for social change and the arena in which they can be a permanent presence in political conflict. Hence the enormous need for genuinely political, grassroots municipal movements, united confederally, that are anchored in abiding and democratic institutions that can be evolved into truly libertarian ones.

Life would indeed be marvelous, if not miraculous, if we were born with all the training, literacy, skills, and mental equipment we need to practice a profession or vocation. Alas, we must go though the toil of acquiring these abilities, a toil that requires struggle, confrontation, education, and development. It is very unlikely that a radical municipalist approach is meaningful at all merely as an easy means for institutional change. It must be fought for if it is to be cherished, just as the fight for a free society must itself be as liberating and self-transforming as the existence of a free society.

The municipality is a potential time-bomb. To create local networks and try to transform municipal institutions that replicate the state is to pick up a historic challenge—a truly political one—that has existed for centuries. New social movements are foundering today for want of a political perspective that will bring them into the public arena, hence the ease with which they slip into parliamentarism. Historically, libertarian theory has always focused on the free municipality that was to provide the cellular tissue for a new society. To ignore the potential of this free municipality because it is not yet free is to bypass a slumbering domain of politics that could give lived meaning to the great libertarian demand: a commune of communes. For in these municipal institutions and the changes that we can make in their structure—turning them more and more into a new public sphere—lies the *abiding* institutional basis for a grassroots dual power, a grassroots concept of citizenship, and municipalized economic systems that can be counterposed to the growing power of the centralized nation-state and centralized economic corporations.

November, 1989

Workers and the Peace Movement

A vigorous attempt is being made, these days, to create a sense of guilt among peace activists who blockade weapons-producing plants. Such activists, we are told, tend to be politically and ideologically "counterproductive." They "alienate" workers who are "compelled" by their material needs and responsibilities to produce weapons. These workers, it is argued, are obliged to make weapons by a system they never created. They should be spared the problem of encountering peace activists who obstruct their access to their plant—or to use a more poignant word, their "jobs."

This argument, advanced most forcefully by traditional Marxian socialists, tends to acquire a particularly odious form when a "class analysis" is tagged on to the issue. Peace activists who blockade weapons-producing plants are snidely depicted as "middle-class" elements who, to use the language of the former socialist mayor of Burlington, Vermont, are either free to "choose" their own jobs and lifestyles or are sufficiently well endowed financially to be spared the need to hold jobs of any kind. The workers, by contrast, are depicted as "victims" of the war machine who cannot afford the "luxury," if they wish to "survive," of selecting their own forms of work and ways of life. They should thus be shown the deference that righteous people accord to victims of injustice of any kind.

This comparison of the "middle-class peacenik" with the sturdy "proletarian" weapons-producer has particularly sinister implications. Merely on the face of things, the socialists who advance these arguments actually fuel feelings of confrontation between two groups of people who, in fact, they should try to reconcile. By defending "jobs" as such without clearly demarcating between work which results in weapons production and work which results in useful goods, they often provide rationalizations for responding passively to current policies of military expansion and militarization. The ethical aspects of work in this society ceases to be an issue in dealing with workers as human beings. Indeed, workers who may be uneasy about the weapons they are producing can be tipped by this kind of class-based rhetoric into dealing with their image of their work along "class" [read: chauvinistic] lines and comfortably ignore the profoundly moral issues raised by a blockade.

What is equally reprehensible, the moral issues raised by a blockade are eclipsed by strictly economic ones. Education gives way to deference to workers whose jingoism and prejudices should clearly be brought into question. Many peace activists, in turn, are not "middle-class" or privileged people. Guided by deeply ethical concerns for injustice and war, they often elect to live more austere lives than conventional middle-class people and "proletarians." Their decision is by no means a luxury; indeed, it often involves very demanding sacrifices and hard work. To call an ethical decision a "luxury" reflects a Neanderthal mentality toward morality that has nothing in common with the highest ideals of human freedom and conscience.

Granted that peace activists should try to inform workers in so-called "defense plants" of the effects of their jobs. Granted, too, that they should advance vocational alternatives to the production of weapons and perhaps no less significant, advance social alternatives to the *inequitable* distribution of useful goods that prevails today. This is an aspect of many "peace budgets," with their emphasis on jobs, that has been woefully ignored. But it is demagogic in the extreme and morally odious to dismiss ethical considerations of justice and conscience as a "luxury" and economic considerations of vocational options and often dubious forms of consumption as a "necessity."

The *need* to be moral—a truly *human* necessity—is degraded by a cynical sense of expediency that is no different in principle than Reagan's Cold War saber-rattling in the name of the "national interest"; the goals may be different but the way of thinking is uncannily similar—and the way of thinking may easily shape the goals themselves, as past experiences have shown.

Some sections of the "New Left" of the late sixties, using this amoral sense of expediency, literally destroyed their credibility by making our enemies' enemies into our "friends" and extolling totalitarians like Mao Tse-tung, Ho Chi Minh, and even the late and unlamented Joseph Stalin, as their "revolutionary" guides. The moral revulsion this produced among millions of people was more than understandable; it was literally admirable. Later, people like Joan Baez were almost hounded out of "The Movement" when, out of moral outrage, she condemned the horrible massacres conflicting socialist camps were perpetrating in Cambodia and southeast Asia generally. Susan Sontag, outraged by the socialist suppression of Polish Solidarity, was roundly condemned by Communist hacks and their supporters as a "red-baiter" for rightly condemning the totalitarianism of the Eastern zone as fascistic.

This kind of mentality—of sleazy expediency at the expense of principle and moral integrity—is merely the dogmatic "Left's" McCarthyism in reverse. It places the manipulative lie above an honest regard for facts and decency. It extols power politics over humanistic ideals, economic calculation over moral probity, rhetoric which always implies that people are mere instruments to be used for political ends over a regard for human beings as ends in themselves.

Actually, this mentality is simply Cold War demagoguery carried from the diplomatic summits of international affairs to the gates of a weapons-producing factory. Cheap prejudices about the "luxury" of morality are used to win constituencies that have yet to define their own ethical standards about producing weapons which may slaughter countless numbers of human beings who live, for the most part, in the Third World. If one wants to talk about calculation or, to use a typical Marxist euphemism, "tactics," when is a blockade "permissible" and when is it "counterproductive"? When a conglomerate decides to produce nerve gas instead of Gatling guns? When a utility decides to build a breeder reactor instead of a machine-gun plant? When a multinational corporation decides to manufacture missile-guidance systems instead of erecting a nuclear reactor? Note that in all such cases, construction workers could vociferously complain that they are being denied access to "jobs" because peace activists are blocking the roads to construction sites and plant operators could make a similar claim about their own "jobs" after an installation has been built.

What it all comes down to is clear enough: once one discards moral criteria for "tactical" calculation in assessing the validity of a war-plant blockade, principle is surrendered to expediency as a guide to our actions and ideas. The ultimate effect of such unprincipled thinking and behavior can be utterly devastating. When one's enemies' enemies mechanically become one's friends, when a specious "class analysis" and abstract social categories supplant a reasoned and ethical consideration of reality, when McCarthyism is played on a piano with a red keyboard instead of a red-white-and-blue one, people themselves are degraded into mindless robots whose decisions could just as well be made by computers as by the rhetoricians of the "Left." They can then be induced to dance to tunes orchestrated by the blocs of the Cold War rather than act with the spirit of defiance, independence, dignity, and morality of true human beings. Their very humanity, in effect, is violated in the name of achieving a humane society—a consequence that begins as the Stalins, Maos, and Hos have shown to alter the very meaning of the word "humane," not to speak of the word "freedom."

The Myth of the Proletariat

The truth is that the Marxist critics of war-plant blockaders *do* assign a highly privileged status to a distinct group of people, one that is scarcely guided by an honest evaluation of the "rights-and-wrongs" of a specific issue. Far from looking at such issues on their own terms, they view them through an ideological prism that assigns a sweeping historical privilege to "The Proletariat." It is this purely dogmatic prejudgment and bias that almost consistently motivates them to condemn any critics of "The Proletariat," with little regard for the validity of the issues in themselves.

A "class analysis," so hopelessly ossified by ideology and mystical faith, becomes an eerie forcefield of theoretical categories that totally immunizes its acolytes from any contact with history and reality. "The Proletariat," conceived as a category, in fact, devours the very proletarians who presumably compose the class which is imbued with this lofty status, a notion that made it possible for Marxists like Lenin and Trotsky to suppress authentic grassroots working-class movements for soviet democracy in Russia in the name of the "historical" role of "The Proletariat" as a category. Individual workers, to be sure, can be strike-breakers, labor spies, wife-beaters, opponents of unions, red-hailers, racists, and even members of the KKK, like everyone else in all classes of society. At times, the number of such reprobates can be so enormous that to overlook their behavior and views is ideological myopia that verges on sheer blindness. But like a flag that is tattered and torn, "The Proletariat" marches forward in modern history as a hegemonic stratum, destined perhaps in spite of itself to change the world radically and produce a free and humane society.

This theory has been imputed to Marx, his followers, and so-called "nonauthoritarian" socialists like anarcho-syndicalists, a theory that has been propagated for over a century with very dismal consequences. Today, it is advanced mainly by dogmatic Marxist and syndicalist leftovers from the 1930s, although the cartoon-myth of the "sturdy worker" still captivates many people in liberal and left-of-center circles. That there have been perceptive debates over the historical role of "The Proletariat" in more informed socialist circles of all kinds seems to have largely eluded the orthodox acolytes of the Communist Party, assorted Trotskyite and Maoist groups, and the devout remnants of American syndicalism. The theory was grossly unsatisfactory while it was in Marx's own custody. At present, it has been grossly vulgarized by certain Marxists and anarchists, few of whom take the pains to read the works of their "founding fathers" directly, much less reflect on their own ideas with minimal independence of mind.

To understand what is so gravely flawed about the theory of "proletarian hegemony"—and the confusion its failings have created in peace, ecological, and feminist movements throughout America and Europe—we must briefly examine some of its theoretical premises. Perhaps the most obvious flaw originates in the way Marx tried to reason out the transition from capitalism to socialism. Just as Marx thought of a "proletarian revolution" largely in terms of the Great French Revolution, and even planned in his youth to write a history of the latter, so he thought of the transition from capitalism to socialism largely in terms of the development of capitalism out of feudalism. The bourgeoisie had undermined feudalism by growing and expanding within it—in the medieval communes of the West and along the trade routes that gradually eroded manorial self-sufficiency by a rambunctious and vigorous market economy. Like a fetus in the mother's womb, it developed to a point where it became sufficiently mature to come into its own right. In the case of the bourgeoisie, capitalism simply destroyed feudalism, not only economically during the late Middle Ages, but politically in the revolutionary era of the eighteenth and nineteenth centuries.

This historical "scenario" is crucial to an understanding of Marxism. It was easy for Marx to draw a corresponding parallel between the bourgeoisie and feudalism on the one hand and the proletariat and capitalism on the other. In Marx's view, the proletariat was a fetus that would grow in the womb of the established society, just as the bourgeoisie had developed centuries earlier. It too would reach a degree of development that would compel it to destroy its own "parent"—the bourgeoisie and capitalism. Like the bourgeoisie in feudal times, the proletariat was the harbinger of a radically new society—socialism, which in turn would develop into a classless, Stateless, and radically new distributive economy called communism in which work would be performed according to one's ability and goods would be distributed according to one's needs. Communism, in effect, would resolve the infamous "social question" of human exploitation, material scarcity, class privilege, private property, and State domination.

Symmetrical, even artistically balanced, as this parallelism between two concepts of transition may seem, it is deeply flawed by inconsistencies. The bourgeoisie—by no means to be confused with guild craftpersons—was a truly alien force in the feudal system. It developed an economic identity that was not only non-feudal but anti-feudal. It required "free labor" that was not tied, like serfs, to the nobility's landed estates, and it needed a free market that was not restricted by guild regulations and a self-sufficient manorial economy. Initially, it also needed a highly centralized State to counteract feudal decentralization, a State that could create a well-policed nation that would safeguard its merchants from arbitrary feudal lords and their private armies who robbed trade caravans, tolled them when they crossed from one feudal domain to another, and freely issued their own debased coinage which traders were often obliged to accept at the peril of their lives. Perhaps more significantly, the bourgeoisie not only developed within feudalism as an alien entity, hostile from the outset to feudal social relationships and institutions; it literally replaced the feudal economic system by commercial and industrial capitalism long before it replaced the feudal political system by a republican nation-state that advanced its specific interests at home and abroad. Capitalism, in effect, already predominated economically in strategic areas of Europe long before it used the English, American, and French revolutions to predominate politically in much of the Western world.

It is vital to recognize that the proletariat has never gained this kind of economic—much less political—predominance within capitalism. Indeed, what is more disconcerting: it is doubtful, on close analysis, if we can even regard the proletariat as an *alien* element within capitalism, much less the fetus that will yield a new society. Marx's theoretical parallelism between the transition from feudalism to capitalism and from capitalism to socialism is anemically superficial at best and grossly misleading at worst.

Lacking economic power in the form of a socialist economy—a concept that itself requires careful and critical elucidation—the proletariat must first "seize" political power before it can transform society in any way. It makes no difference whether one describes this "seizure of power" in economic terms or political ones, say through a general strike in which workers take over the factories or a political insurrection in which workers take over the State. Either way one views it, be it syndicalist or socialist, the confrontation is political in character because the working class must expropriate something which it does not possess and, in so doing, must confront the full institutional and military array of a political entity called the State. Spanish anarchism foundered on the belief that by taking over the factories in Catalonia, it controlled Catalan society. In the meantime, the Catalan bourgeoisie marshaled its own forces within the arena of the State and easily reclaimed its "expropriated" economy with no resistance from the Spanish proletariat as a whole and merely a futile defensive action by the Barcelona workers in May 1937.

The primary political act of expropriation in the *full* sense of the term, not mere factory takeovers, presupposes a degree of political consciousness and theoretical insight that a traditional working class has rarely, if ever, exhibited. Workers may be militant, but they are not necessarily revolutionary—even in what Marxists and anarchists call "revolutionary situations." Working-class militancy generally occurs within an economic and political framework of social reform not social revolution. In fact, all the great "proletarian revolutions" the Marxists and anarchists celebrate such as the Paris Commune of 1871, the Russian Revolution of 1917, and the Spanish Revolution of 1936, were made not by a hereditary working class whose parents had been socialized in factories. They were made by preindustrial artisans (Paris in 1871) or peasants (Russia and Spain) who had recently flooded urban industrial centers because they lived at a near-starvation level on the land. These "proletarian revolutions," in effect, were as much the result of cultural dislocation from a seasonal world governed by nature to an industrial world governed by time-clocks as they were the result of conflicts between wage-labor and capital. In the one case where they succeeded (Russia, 1917), they were reinforced by soldiers who were merely peasants in uniform, without whom the working-class Red Guards would have been powerless against the old order.

Where hereditary workers did "revolt," such as in Germany in

1918–19, they were completely controlled by reformist Socialists who were as hostile to revolutionary action as the bourgeoisie. It is notable that such outstanding Marxist revolutionaries as Rosa Luxemburg and Karl Leibknecht could never gain enough votes from German workers to become delegates to the Congress of Workers' Councils which socialists had convened in early 1919. These distinguished revolutionaries, later to be killed by the military after a fruitless uprising by their followers a few weeks later, were reduced to the indignity of heckling the Congress from the balcony.

To claim that proletarian revolutions have such an abysmal history because they require a conscious and committed revolutionary party to "lead" them raises more questions than it answers. In the first place, it fails to explain why such parties, when they do arise, rarely gain the confidence of the very workers in whose "historic" interests they profess to speak. One has only to look at such industrially advanced countries as the United States, England, Germany, and France to see how futile has been the career of any genuinely radical, not to speak of revolutionary, party. Eugene V. Debs's Socialist Party in America, the radical factions of the British Labour Party, Luxemburg and Leibknecht's Spartacus League, and the early French Communist Party were largely ignored by the working classes in their respective countries. The workers who entered these parties did so for a wide diversity of reasons and never constituted a majority of their class. Even within the Spanish anarcho-syndicalist unions which collected sizable radical working-class minorities, the more forthright of their leaders acknowledged that less than a quarter were truly committed anarchists, despite the overall militancy of their members.

Where so-called "workers' parties" have succeeded in carrying out a revolution, notably in Russia, they clearly betrayed their programs within a decade after coming to power. The ultimate heir of the Russian Revolution was Joseph Stalin, perhaps the most terrifying mass murderer and certainly the most vicious counterrevolutionary in our century. To adduce the dubious success of "revolutions" by peasants in Global South countries as evidence of successful "workers' parties" verges on sheer idiocy. Not only do industrial workers, the classical "revolutionary agent" of socialist theory, constitute a very small minority of Global South populations, but the revolutions in colonial countries, despite the rhetoric of their "Marxist" leaders, are too often patently nationalistic in character, parochial in outlook, and authoritarian in their results.

The particular brands of hatred that riddle Soviet-Chinese relations, southeast Asian factions, and Central American guerrilla bands are not simply national in character; they are overtly racist and draw from the worst sewers of Euro-American ethnic antagonisms. One can vehemently oppose American intervention in the Global South and still take critical cognizance of "revolutions" that have been thoroughly permeated by racist, sexist, bureaucratic, and authoritarian features.

We can remain at the mere level of obvious facts and skin-deep analyses by holding on to the traditional explanations of why "proletarian revolutions" have failed so consistently by speaking of betrayals, "opportunism," and the "co-optation" of workers' parties. This does not tell us why such "betrayals," "opportunism," and "co-optation" occur so consistently. Or we can try, as Marx might have done had he lived into our own time, to probe more deeply into basic problems that exist in the very concept of "proletarian revolution" and ask strategic questions about "The Proletariat" itself and the kind of class politics and organizations that are spawned by the traditional Marxist and syndicalist visions of the class as such.

To put these basic problems quite bluntly: is "The Proletariat" an "embryo" within capitalism that is "destined" by history to play the "revolutionary role" that capitalism played in its conflict with feudalism? Can it rise as a class—and I speak of a class, not of working people who are subject to problems and issues that face humanity as a whole—beyond the conditions that capitalism imposes upon it as a class? Is "The Proletariat," far from being an "embryo" of a new, alien social being, an organ that is as much a part of capitalist society as any organ in a human body? Have Marxists or syndicalists any more right to assign a historically privileged function to "The Proletariat" than liberals do, say, to "The Middle Class"?

There is ample evidence that Marx himself was more deeply

aware of the flaws in his transitional parallelisms between the historical role of the bourgeoisie and the historical role of the proletariat than any of his acolytes. Unfortunately, he tried to erase these flaws with explanations that on close inspection reinforce and deepen them. Clearly aware that "The Proletariat" could not build a socialist economy—much less a socialist society—within capitalism the way the bourgeoisie built a capitalist economy within feudalism, Marx began to look to the capitalist economy *itself* as the source of a revolutionary movement led by working-class organizations. His "immiseration theory" is the most famous case in point. Capitalism, he was to argue, was destined to impoverish the workers to a point where they would be compelled by sheer desperation to rise up in social revolution and overthrow the bourgeoisie. That this vision has yielded no socialist or even revolutionary consciousness in the industrial heartland of the world even in periods of grave economic crisis has haunted radical social theory for generations.

It is time to acknowledge that a worsening of economic conditions does not inspire a revolutionary consciousness, much less a socialist one, but fosters demands for reform within the capitalist system and, perhaps more poignantly, resignation and slavish attempts to accommodate to the vocational demands of society. The bourgeoisie, in turn, has become sufficiently sophisticated over the past century-and-a-half to manipulate economic conditions and interplay them with political alternatives such that it successfully allays discontent when it becomes too serious or diverts it through such mechanisms as jingoism, Cold War ideologies, racism, and regionalism, to cite only the major channels for siphoning off popular anger.

But no less significant than Marx's "immiseration theory" is the organizing principle Marx was to extend from the bourgeoisie—indeed, from the capitalist mode of production—to "The Proletariat." In Marxian theory, capitalism was a "revolutionary society because it was inherently secular and rationalistic. The "bourgeois" Enlightenment, in Marx's view, like trade and industry—and the calculating sensibility they required—"demystified" the world and exposed the economic base of society in all its egoistic nakedness.

The Communist Manifesto and many passages in the *Gründrisse* reveal an extraordinary admiration for the rationalistic side of capitalism—its aggressive attempt to bring all the "forces of nature" under human control, to organize the labor process efficiently, to advance technology without regard to tradition, myth, and presumably "reactionary" attempts by "outlived" classes to preserve "archaic" lifeways and technical operations.

To Marx, "The Proletariat" was the heir to the bourgeoisie's secularization and rationalization of the world. It was destined to continue the work of capitalism along more positive and humanistic lines while still retaining the discipline and authoritarian character of capitalist production and industry. This destiny and "historical task" did not emerge from mere speculation. It was formed in the crucible of the capitalist mode of production itself—in the factory and as a result of the inexorable demands of industrial production. The working class, to Marx, was "a class always increasing in numbers, and disciplined, united, organized by the very mechanism of the process of capitalist production itself." The working class, in effect, was the human raw material and embodiment of capitalism's historically "rational" role—rational in its unity, its discipline, its organization—and hence as much a "mechanism" for revolutionary social change as the capitalist mode of production in the long historical movement out of a mythic and mystified preindustrial world. Industry would transform the nonproletarian oppressed with its mystical shibboleths, irrationalities, and amorphous qualities into a hegemonic "Proletariat" that would move toward secularism and rationalism without the "cash nexus" that unites capitalism.

Alluring as this notion may be, it is completely deceptive. Capitalism assuredly does rationalize its labor-force and the process of production. But it rationalizes them *hierarchically*. No less than the hapless feudal serf or the ancient slave, the proletariat is indoctrinated from birth, schooled, and finally unconsciously habituated on the job site to an all-encompassing system of obedience and command. If the factory and the capitalist mode of production have any effect upon the workers, this effect is negative. Both internally and externally—psychologically and economically—"The Proletariat" is

brutally shaped more so than artisans or agriculturists by the hierarchical discipline of industry and the ever-confining domineering
role of an increasingly rationalized work environment. The ticking
of the time-clock dominates not only the worker's day but invades
his or her very psyche to yield a carefully machined object that is
imbued with the mechanical routine of the factory. The worker is not
only "disciplined, united, and organized by the very mechanism of
the capitalist mode of production"; he or she is psychologically, economically, and vocationally absorbed into it and becomes integrally
part of it.

To Marx, this integration, even psychologically, was not an
undesirable attribute of "The Proletariat." But in retrospect, it is now
clear that the harsh reality of the factory shapes the destiny of the
worker as a class being—and it does so in a manner that is extremely
reactionary, indeed nonrevolutionary. Command, obedience, and
rationalization become "The Proletariat's" most significant *class*
traits and form its image of organization and society. The German
Social Democrats who were the target of Lenin's most venomous
language such as "betrayers" and "renegades" actually betrayed no
one and reneged on nothing insofar as they were spokesmen for the
majority of the German working class in 1918–19. They unerringly
spoke for a class that had been trained throughout life by the capitalist mode of production to accept the core attributes of capitalism
and thus capitalism itself with its offerings. As Rosa Luxemburg so
clearly understood, they did not want a revolution when they were
summoned to rise up in January 1919, under the auspices of the ill-
fated Spartacus League or the newly formed German Communist
Party—and their leaders, particularly Ebert, Scheidemann, and
Noske, assured them that they would not have one. They wanted an
orderly democratic republic—as orderly as their factories and as
democratic as their unions. They wanted a disciplined society—as
disciplined as their jobs. They wanted a united Germany under the
Weimar Constitution—as united as their workshops and assembly
lines. They were accustomed to obedience and command on the job,
and they obeyed Ebert, Scheidemann, and Noske when they doffed
their uniforms after the war and returned to work.

The Social Democratic Party never lost the allegiance of the German working class and its majority never followed the radical tendencies in German socialism. History allows for no other verdict than the fact that, except where peasants were placed in a thoroughly destabilizing forcefield of changing lifeways from the agricultural to the industrial. The majority of the working class consistently followed the lead of reformist labor parties—even in so-called "revolutionary situations" where major social changes seemed imminent.

Rather than contrive a maze of convoluted explanations for this historical verdict, we would do well to face reality directly and candidly. "The Proletariat" is no more a hegemonic revolutionary class—*qua class*—than the ancient slaves and plebians or the medieval serfs. Just as the latter lived within classical or medieval society and faded away with it, so "The Proletariat" as a *class* is inexorably part of capitalist society and its destiny is tied to the development of the bourgeoisie. That the historical-transitional parallel Marx created between the bourgeoisie and the proletariat was riddled by crucial inconsistencies in his own mind is evidenced by the extent to which he relied on capitalism itself to produce the economic and ideological conditions for its transcendence into socialism— conditions which we now know are utterly specious. Indeed, to the extent that capitalism has developed technologically and politically; to the degree that it commands the "forces of nature"; to the extent that it has concentrated the powers of technics and the State—to that degree its command of humanity's destiny with a technics of war and with institutions of control raises more *formidable* obstacles in the way of radical social change and even the formation of a radical consciousness than any society in history.

Finally, it is fair to ask if *any* class within class society can give rise to the classless society we call "socialism" or "anarchism." The bourgeoisie, which emerged within feudalism, did not produce a classless society, however alien it was to the manorial economy of the feudal world. The "historical task" Marx assigned to "The Proletariat" is even more formidable. To expect a class whose mentality is shaped by the lifeways, habits, values, culture, and hierarchical interdependence of another class—the bourgeoisie and its industrial system—to transform society as a class with its own class interests, indeed to "negate" in the larger interests of humanity (interests which have yet to become part of its own) is preposterous. "The Proletariat" is neither better nor worse than the society and economy that produced it—and it is a class that, in Marx's own formula, reproduces itself daily in its service to the bourgeoise, *culturally* as well as economically. In contrast to the early bourgeoisie, it has never existed apart from its social milieu or in revolutionary tension with it.[1] It has merely demanded its "rightful place" in capitalist society, and, like the peasantry that once acquired land as a result of the "bourgeois revolutions," it has largely acquired it—that is, insofar as it is needful to the operation of the capitalist system. It will never participate in revolutionary social change until it transcends its class being as part of capitalism and like the early bourgeoisie becomes part of another social being that is as alien and hostile to capitalism as the bourgeoisie was alien and hostile to feudalism.

Theoretical Reconstruction

Either we will undertake a ruthless reconstruction of radical social theory and analyses in the face of these considerations or we will remain the mindless victims of dogmas that we have inherited from an era that is long-gone and now plays a completely obfuscatory, indeed, reactionary role in social consciousness.

So far as "The Proletariat" is concerned, this reified and objectified category must be demystified. Its role in radical theory and practice must be placed in a real-life perspective, not in a transcendental Kantian realm. "The Proletariat," or, for that matter, "The Peasantry" and "The Petty Bourgeoisie," are historically very similar. "The Proletariat" reproduces and supports the wage system insofar as it is necessary as wage labor, the counterpart of capital. Its wages as "variable capital" are no less capital than "constant capital" (i.e., the factory and raw materials needed for capitalist production). Both are tainted by the conditions that are necessary for the existence of capitalism. These relationships are not mere abstractions. Exploited as "variable capital" may be, this exploitation is part of its very identity in the form of a commodity called labor power. This commodity as embodied in the worker is the cultural carrier of industrial and social hierarchy. It is present in the worker's demeanor, forms of organization, parties, cultural preferences, attitudes toward deviance of any kind, tastes, treatment of the opposite sex, children, ethnic groups, the nation-state, and even the class which oppresses it. Alienated as the worker may be, his or her alienation perpetuates the process of production by rendering it competitive, egoistic, and, more decisively, by rendering the worker dulled and anesthetized to deadening industrial processes to an extent that even peasants-in-overalls find it intolerable, hence the radical role of the worker-peasant in Russia, Spain, Italy, and France, the authentic homelands of "proletarian" insurrections.

The fetishization of "The Proletariat" has its exact mystical counterpart in the fetishization of the commodity and the fetishization of needs. To the degree that radical theory objectifies and fetishizes the proletariat, to that degree it is victim to the myths of a class within capitalism that will emancipate humanity from class society as such, of a class integral to capitalism that parallels the function of a class that was alien to feudalism, of a class regimented by industry that will magically dissolve its hierarchical habits, of a class integrated by mass culture that will be heir to the best in human culture.

Let us agree that no radical social change is possible without the support and initiative of working people—or, for that matter, of technicians, professionals, soldiers, women, ethnic groups, youth, the elderly, and the solidarity of the oppressed on a worldwide scale. But no radical change is possible unless "The Proletariat" transcends its suffocating class being and becomes a revolutionary *human* being. This transcendence involves the erosion of the worker as a class being, the acculturation of what is still a class into a people with a conscious sense of the public good in contrast to class interest. It is ironic that bourgeois society, which more so than any other society is built around contract and interest, presupposes precisely a sense of class interest within "The Proletariat" that socialists and syndicalists have nurtured as "intrinsically" revolutionary.

The Marxian myth that the working class will be driven by its own crass economic interests to overthrow capitalism is the ideological mystification of a real fact in bourgeois society: notably, that the crass economic interests of "The Proletariat" are an objective precondition for the existence of capitalism because the capitalist mode of production presupposes a denial of an ethics of the public good in favor of a *means-ends rationalism* of personal and class interest with the contractual bonds to specify these interests like a bill of lading. Insofar as "The Proletariat" embodies this interest, it becomes party to a shared negotiation-process of "give-and-take" that daily keeps the capitalist mode of production and all its cultural paraphernalia alive. Through its unions and parties, "The Proletariat" becomes a negotiator within capitalism, not the famous "grave digger" of capitalism, a role which Marx imputed to it historically.

We arrive at the quixotic fact that "The Proletariat" can play a socially transformative role only insofar as it ceases to be a mere class and arrives at the selfhood and self-consciousness that renders "it" a constellation of human beings—human beings who can generate the self-activity involved in self-management.[2] The word "self"—individuation, consciousness, rationality, and the fulfillment of one's potentialities to function as a citizen in a free public sphere—is the common denominator of these human traits. In the broadest sense, what I am saying is that working people become radical people *despite* the fact that they "work" rather than because they work. They do so because they can be motivated to transcend their mere being as workers. That is to say, they become radicalized only insofar as they can become ethical beings and function primarily on an ethical rather than economic level of their existence. To appeal to workers as human beings rather than as job-holders, to appeal to their conscience rather than their material needs, to appeal to their sense of right and wrong rather than their "interests"—all of these appeals form the indispensable means for transforming "The Proletariat" into human, ethically motivated individuals who can challenge the entire constellation of hierarchy, domination, and unfreedom, a constellation that brings what we call "civilization" itself into question, not only capitalism.

The radical establishment in all its forms—the "Left" as it has been called from its origins in the Jacobin era of the French Revolution—fosters the imagery of the worker as a job-holder, a creature of raw "interest," an embodiment of brute need over his or her human traits as a person of conscience and rationality. It gives social and psychological priority to the worker precisely where he or she is most co-joined to capitalism and most debased as a human being— at the job site. It mystifies the factory arena precisely by exalting it as the "historic" locus of confrontation: where wage labor encounters capital. Whoever seeks to transcend this locus by appealing to the worker's humanity, conscience, and personality—as a being that is richer than a Marxian category—is condemned as a "class collaborationist." Whoever intrudes upon the negotiations between wage labor and capital where the worker is brought into complicity with his or her own degradation—a degradation reinforced ideologically by the Marxists and the syndicalists—is condemned by the "Left" and "Right" alike as an obstructionist.

Yet it is precisely when peace activists, feminists, ethnic groups, gay people, environmentalists, and countercultural folk pose the ethical question of complicity with militarism (including the militarism of the factory), a machismo sensibility, racist attitudes, issues of sexual preference, the degradation of the natural world, and the hierarchical schooling and debased lifeways of "The Proletariat" that workers have the possibility of transcending their class and mystically reified class character. The ideological challenge of a peace-activist blockade at a "defense plant" stands on a qualitatively higher level than the attempt of socialists and syndicalists to win "The Proletariat's" support by condemning such actions and challenges. Whether knowingly or not, peace activists raise the ethical issues that alone can separate the human being within the worker from his or her economic being. They raise the moral dilemmas that alone can turn the worker from a "factory hand" into a person of conscience and nourish the human spark that exists in every working person. They raise the crucial issue of work in this society—its meaning, goals, and creativity as distinguished from its debasing psychological effects, its role in the service of social and industrial hierarchy, its function in producing profit and lethal means of destruction.

That peace activists are abused at the factory gate and often violently assaulted is evidence that deep prejudices are being challenged. Such prejudices do not die easily—nor should we ever expect that they will do so. Workers who react to challenge are in a deep sense *acting* or *being acted upon*. Their complacency will never be shaken by "Leftists" who justify these prejudices in the working class and deepen them for manipulative purposes. Here, the "Left" plays a particularly reactionary and sinister role by introducing the prejudices of uninformed workers into the peace movement itself, often for purely opportunistic and immoral reasons. Whether the war machine will cease to operate because of peace blockades is not at issue. What counts is that the prejudices of ages, from the work ethic to the notion of "interest," are shaken, indeed that consciences are deepened by rational challenges and ethical behavior.

It will not be blockades, to be sure, that will solely affect the conscience of working people and the operations of weapons plants. Larger processes are now at work which are eroding "The Proletariat" as a class with the same long-range effects that ultimately yielded the virtual disappearance of the peasantry and small farmer. Robotics, cybernetics, and Japanese management techniques may leave us with a working class so numerically small and culturally warped that it will no longer be recognizable by its ideological acolytes on the "Left." We are increasingly surrounded by the living dead in the industrial world, despite the predictions of Marx and Engels. The smokestack industries and traditional assembly lines are largely closed down for good, a development whose full outcome is beyond the purview of the present. The "historic" role of "The Proletariat" is becoming moot; the class itself is changing radically in structure and numbers—more radically than it ever changed society.

These sweeping changes in the traditional class structure of bourgeois society—and they affect *all* strata that belong to the past— are crucial. Out of the decaying proletariat, middle classes, and agrarian classes, a new phenomenon is emerging. The old social pool called the "people" is being restored in the tension between past and future, a classless "class" like the *sans culottes* composed of economically, culturally, and technologically displaced persons. The unstable men, gender-conscious women, deprived minorities, the aged who have no sense of status, the youth who have no sense of future, the cultural dissidents—all lack a feeling of place in the society or a stake in its existence. To call them a "new proletariat" is to render the word meaningless in the Marxian sense of a class "disciplined" and "united" by the capitalist process of production. They constitute what the syndicalists brightly call "marginalia" that, like all historical fringes, threatens to move increasingly to the center of society and become a source of endless ferment and deinstitutionalization.

It is this classless "class" that is most receptive to the larger demands and issues posed by a decomposing traditional society— the demands and issues of peace, sexual freedom, and ecologically, socially, and culturally meaningful alternatives to the status quo. And it is from this classless "class" that the peace movement can draw its most forceful and compelling activists. Unless war and ecological breakdown bring the entire human adventure to an end, this classless "class" can only exist for a limited period of time—between the breakdown of a relatively democratic society and the emergence of a highly centralized, well policed, and militarized one. If it does not find its "program" and its own organizational forms within a generation or two, it will be eliminated in one way or another like earlier transitional strata by a totalitarian system that will close all the doors to social change.

A second consideration that has eluded socialists and syndicalists is that workers live in communities, not merely in factories. In their neighborhoods and towns, they tend to reveal significantly different traits from those that are fostered at their job sites. They are men who fear to become cannon fodder, women who feel degraded by their male-dominated milieu, ethnic people who carry the added burden of racial discrimination, gay people who are obliged to conceal their sexual preferences, uneasy residents of polluted environments, and politically disempowered individuals who seek more control over their daily lives. They are parents, youth, sisters, brothers, concerned or fearful people who seek the rights of a long-denied citizenship.

To many of them, their jobs are merely means to an end—often as decent, peaceful, and, yes, idealistic as those of the peace activists who blockade their plants. They can be reached as neighbors more readily than as metaphysical beings who comprise an abstract category called "The Proletariat." Their homes, families, friends, and life-problems mean more to them than their class "interests," particularly as these are mystically interpreted by the "Left." Their human locus is the community in which they live, not the factory in which they work. In the shifting world of disintegrating classes, they can be more open to serious discussion as people than as "proletarians."

Socialism and syndicalism, with its bourgeois mentality of economic reductionism, denies them their humanhood. By contrast, peace activists build on the personality and individuality that distinguished their personhood from their workerhood, their humanity from their economic debasement, their sense of citizenship from their enfeeblement as taxpayers. They reach out to the potential revolutionary who is papered over by the supine reformist. Behind socialism and syndicalism lie nearly two centuries of abysmal failure and an ever deepening dogmatism. Their acolytes can only yelp—and feel guilt—as though they live in remorse for their stillborn world and interpretation of reality. Many prejudices have yet to be broken, not only among workers but among the elements who foster their biases and tilt them toward accommodation to a ruinous society whose future is beyond redemption. Civil disobedience cannot be abated and its supporters cannot defer to the pain that comes with the process of dissolving biases and dogmas unless the peace movement decides no longer to obstruct the way to an apocalyptic devastation of nature and society. Once the impediment to this catastrophic course is lifted, either by guilt or uncertainty, the end of this planet will be a certainty—and there will be no compensating resurrection of humanity and the human spirit.

July, 1983

An Appeal for Social and Ecological Sanity

I

We may well be approaching a crucial juncture in our development that confronts us with a historic choice: whether we will follow an alternative path that yields a humane, rational, and ecological way of life, or a path which will yield the degradation of our species, if not its outright extinction.

If this seems like a reckless overstatement, the cry of an aging alarmist who bore witness to more than half a century of growing ecological and social crises, let us try to assess the kind and scope of the problems that have arisen over the past few decades and the dangers they pose for nearly all complex life-forms, including our own, that have evolved over aeons.

Certainly, the possible outcome of a world thermonuclear war, even with existing weapons, has not been exaggerated by the grimmest of our social and scientific forecasters. Biologically, the human beings who survived such a conflict would have good reason to envy the dead. Firestorms, radioactive fallout, hundreds of millions of decaying bodies, the barren landscape, evaporated or condensed lakes, the debris of cities and towns, the hopelessly ill and disastrously wounded—I leave out the grimmest of all predictions, a "nuclear winter" which will blanket the earth with dust and debris that will shut out the sunlight necessary for life on the planet—all, taken together, give us reason to wonder whether the grossly degraded ecology of the earth would be capable of supporting mammals like ourselves in the years that would lie ahead.

What concerns us in the event of such a biocidal holocaust is not the future of the accursed civilization—with its bloodless gospel of technocracy, egotism, competition, mass culture manipulatory rationalism and, above all, warrior mentality of domination and hierarchy—that will have produced such a terrifying conflict. What concerns us is the future of all remaining complex life-forms as such. For the green world of life which still surrounds us will be replaced by the blackened world of an incinerated biosphere—its atmosphere filled with the stench of the dead, its soil and water polluted by deadly radionuclides, its complex food webs completely shredded with their integrity subverted by disease-bearing organisms and the wild population-explosions of insect infestations. Such a world, all bomb and blast effects aside, would never be one that could sustain the complex plant and animal life forms we know today.

Is such a worldwide thermonuclear war "unthinkable"? While a nuclear conflict between the United States and Russia may not be inevitable, the fluctuation in tensions between the two "superpowers" makes it impossible to answer this question with reassuring certainty.[1]

No longer do we have a "dialogue" between the superpowers that centers on "arms reductions"; rather, we hear "concerns" for "arms equity," a term so loose that it provides no limit to arms expansion. And when American rearmament goals alone soar to astronomical levels, such goals become a major factor in fostering the credo, now widely held, that the "balance of terror" will be advanced by war, rather than the earlier credo that war can be prevented by the "balance of terror." Given this derangement in our formulas for "terror" as an instrument of foreign policy, each side in an oncoming thermonuclear war is obliged to decide when it will strike the "first blow"—not if it will do so—lest the military technology of its opponent become too ascendent to avoid "military defeat."

War, in effect, increasingly becomes a decision that is guided by technological considerations, not only social, economic, or imperial ones. It becomes a matter of "survival" rather than a matter of "victory." Rearmament, especially with nightmarishly exotic weapons, tends not to bring foes together at the peace table. It brings them closer together at potential battlefronts in a spirit of mutual fear and paranoia, not mutual power that balances each against the other.

It is the anemic meaning of the very word "balance" that exacerbates this spirit of mutual fear and paranoia. Consider throughout the Cold War how rhetoric from American leaders was answered in kind by their Russian counterparts. Talk of "limited nuclear wars" has been answered by heightened talk of "imperialist aggression." Each party to this insane babble must make good its fears to justify its threats, to foster and create episodes that validate its claims— hence, prophecy tends to become self-fulfilling. Following out this logic of "terror," American leaders blame the Russians for Central American insurgency while Russian leaders blame Americans for Poland's Solidarity movement. In both cases, these imputations can easily become excuses for widening conflicts either in Latin America or Eastern Europe—conflicts that reflect the rival imperialist schemes of the two superpowers rather than the genuine aspirations of oppressed peoples in El Salvador and Poland. Liberation struggles thus tend to become absorbed into Cold War maneuvers and their authenticity—feared by both superpowers—sacrificed to the strategic needs of blocs on either side of the Iron Curtain.

The need for liberation and antiwar movements to free themselves from any association with either one of these Cold War blocs becomes a massive effort in raising consciousness, a problem that is exacerbated by the fact that these movements are often fighting in mountains and jungles with weapons in hand. Hence, the tragedy of Cuba in the late fifties, of Vietnam in the sixties, and of Nicaragua and El Salvador in the eighties—initially, independent movements that were shrewdly turned into gristmills for cold war politics by the

CIA and its White House administrators. May this not be the fate of Eastern European movements for freedom in their response to KGB persecutions and Kremlin propaganda!

To worsen matters, within each camp—West or East—there are always factions that take every threatening claim seriously, if only for opportunistic reasons of domestic supremacy. These claims serve to enhance the power of self-styled "hawks" over self-styled "doves," each of whom rewords Clausewitz's old maxim to read that foreign policy is merely a means for determining domestic policy. Home control often begins to depend upon the rhetoric—and, ultimately, the action—that is applied to world control.

Given the altered meaning of the word "balance" and the use of foreign policy as an instrument of domestic policy, the new qualitative factor that has entered into the arms race renders all talk about "parity" in a "balance of terror" increasingly meaningless. The very "scale" that was once used to achieve such a "balance of terror" tends to be replaced by a foreign policy based on military action—as witness American intervention in Vietnam and Russian intervention in Afghanistan. Challenge and response between the two world powers turns around attempts to anticipate when and where one power will overstep the bounds that were once defined by the "scale." Hence, any "first strike" risks the possibility of becoming a preemptive strike, a thoroughly neurotic "defensive act" to parry an anticipated "offensive act."

Put bluntly: the "negotiators" at various "peace talks" cease to function as "diplomats," just as Adolf Hitler ceased to practice "diplomacy" at Munich in 1938—an event that was merely a prelude to war. Present-day diplomats have become brutish apemen for whom the outbreak of war serves merely to gauge the fortitude of their opponents—their willingness or capacity to respond. Formulas like "limited nuclear war" deal with biocide as mere "forays" that exist militarily outside the grim possibilities of a worldwide catastrophe. Thermonuclear and biological warfare thus threaten to become the means of achieving a "compromise," rather than "compromise" being a means for avoiding such devastating forms of warfare.

A genocidal strategy is, in effect, trivialized into a mere tactical foray. A sophistication of military technology becomes a factor in initiating anticipatory military actions that traditionally were explicable only in terms of historic economic, geopolitical, and imperial interests. The decision to make war can be determined as much by the rhetoric of a belligerent diplomacy, with its roots in domestic factional conflicts, as by serious plans for conquest. The historic motives that once made for war or peace are being replaced by flippant ones—a downgrading of the horrors of modern war, an excessive exaggeration of military engineering innovations, a purely ideological "battle of words" that can easily become a genuine battle of machines and people. Cinematic satires like *Dr. Strangelove* cease to be mere parodies of a world that can live or die according to the whims of a military commander. They become deadly realities that are more portentous of future events than fears that war will be the result of an "accident" or "nuclear proliferation." Our world rulers, not their lowly subordinates, have trivialized the arms race to such an extent that they can no longer be regarded as the sober custodians of the weapons they have evoked over the past decades.

II

This historic degradation of international relationships is paralleled by a historic degradation of ecological relationships.[2]

The acid rain that has already wholly or partially destroyed half of Germany's forests, and the lumbering of vast rain forests at the rate of an estimated 5 million trees daily,[3] are compelling symbols of the ecological devastation that beleaguers our entire environment. We are no longer talking about the dangers posed by our chemically polluted air, water, food, furnishings, workplaces, and communities. Nor are we talking about the dangers posed by nuclear power plants, acid rain, and the debilitating effects of lifeways that accompany a sedentary, congested, stressful, and highly urbanized world for which our evolution as a species has in no way equipped us. What really concerns us is our destiny as a life form and the future of the biosphere itself.

The possible deaths of vast forests, including the tropical rain forests that girdle the earth, speak to crises that threaten the integrity of our entire ecological fabric. The 1980s opened with climatic changes that are as startling and portentous as the causes that may be producing them. An increasingly dense mist of aerosol droplets seems to be hanging over the Arctic regions which, in the view of many scientists, appears to be warming this geographic cradle of our world's climatic system. Taken together with the widespread deforestation that is still under way, we appear to be subverting the very ecological bases of our seasons, temperature ranges, and the delicate thermostatic systems that factor in every aspect of our weather. Increased solar flares and volcanic activity may provide us with handy excuses for explaining the climatic changes that occurred early in this decade in the United States, but they obscure the long-range seasonal alterations that seem to be affecting the basic biogeochemical cycles of our planet. Major irregularities in temperature, precipitation, periods of aridity or excessive rainfall, and corresponding reactions by plant and animal life portend serious climatic crises within a span of a decade that most forecasters a generation ago had put off for centuries if present pollution rates were to continue.[4]

Industrially and technologically, we are moving at an ever-accelerating pace toward a yawning chasm with our eyes completely blindfolded. From the 1950s onward, we have placed ecological burdens upon our planet that have no precedent in human history. Our impact on our environment has been nothing less than appalling. The problems raised by acid rain alone are striking examples of the innumerable problems that appear everywhere on our planet. The concrete-like clay layers, impervious to almost any kind of plant growth, replacing dynamic soils that once supported lush rain forests remain stark witness to a massive erosion of soil in all regions north and south of our equatorial belt. The equator—a cradle not only of our weather like the ice caps but a highly complex network of animal and plant life—is being denuded to a point where vast areas of the region look like a barren moonscape. We no longer "cut" our forests—that celebrated "renewable resource" for fuel, timber, and paper. We sweep them up like dust with a rapidity and "efficiency" that renders any claims to restorative action mere media-hype.

Our entire planet is thus becoming simplified, not only polluted. Its soil is turning into sand. Its stately forests are rapidly being replaced by tangled weeds and scrub, that is, where vegetation in any complex form can be sustained at all. Its wildlife ebbs and flows on the edge of extinction, dependent largely on whether one or two nations—or governmental administrations—agree that certain sea and land mammals, bird species, or, for that matter, magnificent trees are "worth" rescuing as lucrative items on corporate balance sheets.

With each such loss, humanity, too, loses a portion of its own character structure: its sensitivity toward life as such, including human life, and its rich wealth of sensibility. If we can learn to ignore the destiny of whales and condors—indeed, turn their fate into chic cliches—we can learn to ignore the destiny of Cambodians in Asia, Salvadorans in Central America, Kurds in Syria and Turkey, and, finally, the human beings who people our own communities. If we reach this degree of degradation, we will then become so spiritually denuded that we will be capable of ignoring the terrors of thermonuclear war. Like the biotic ecosystems we have simplified with our lumbering and slaughtering technologies, we will have simplified the psychic ecosystems that give each of us our personal uniqueness. We will have rendered our internal milieu as homogenized and lifeless as our external milieu—and a biocidal war will merely externalize the deep sleep that will have already claimed our spiritual and moral integrity. The process of simplification, even more significantly than pollution, threatens to destroy the restorative powers of nature and humanity—their common ability to efface the forces of destruction and reclaim the planet for life and fecundity. A humanity disempowered of its capacity to change a misbegotten "civilization," ultimately divested of its power to resist, reflects a natural world disempowered of its capacity to reproduce a green and living world.

Technology and science, which staked out such sweeping claims to emancipate humanity from the ages-old burdens of ignorance, superstition, and the resistance of a "stingy" nature, have now been turned against humanity itself—creating new myths of "progress," control, expediency, and efficiency. These new myths threaten to bind our species to an ever-darker fate than the one from which it was presumably rescued. Our medical and chemical armamentarium, perhaps the most celebrated of our technical and scientific achievements, rescues us from microbial diseases only to deliver us, as a result of its industrial applications, to so-called degenerative diseases like cancer. The laboratories which save our lives in childhood with their "miracle drugs" dispose of us in mid-adulthood with their carcinogens. It is as though a generation that has been so successfully pulled from the womb to suffer the bitter travails of life in a highly rationalized and emotionally demanding world must quickly be denied the wisdom of age and the fruits of repose by premature death from the so-called diseases of civilization.

If this verdict reflects the "best" technology and science can deliver in its most humanistic aspects, one wonders what judgment can be rendered for the worst—notably, the results of its patently demonic sphere. Its bombs? Guns? Rockets? Robots? Cybernetic equipment? Chemical synthetics? The fruits of its nuclear and genetic probings? With the possible exception of certain diagnostic tools, surgical techniques, anesthetics, and "magic bullets," no technical or scientific dispensation has done more good for humanity than evil. Technology and science have never blossomed more richly, fully, and fruitfully than in war—the art of killing human beings—with the possible exception of "resource exploitation"—the art of killing nature—often for the purpose of effectively killing more human beings. Here, technology and science join in their most demonic form: the use of nature to destroy people by digging up its hidden stores of uranium for bombs, smelting its ores for guns, applying its laws for obliterating entire cities and ultimately the biosphere itself.

Every genie that has been released from the sacred jars of technology and science emerges with a grateful smile—only to loom over us with a snarl and bared teeth once it has been freed from its confines. Scarcely any technique or fragment of knowledge has been spared from a demonic destiny of killing humankind at a rate that may eventually outpace our species' capacity for procreation. In the twentieth century alone, perhaps 200 million people have been killed either directly or indirectly in wars partly orchestrated by the combined work of the scientist and the engineer. A body of wisdom so inverted that it breeds barbarism rather than civilization, darkness rather than enlightenment, destruction rather than creation, remains even more compelling a challenge to the thinking individual of our times than the concrete problems of thermonuclear immolation.

We of this generation and the past one are not unique in dealing with these issues. Had an Alexander, a Caesar, or a Napoleon—these men whom Hegel knighted "world-historical spirits"—possessed the knowledge of killing that we have today, our species would have suffered its ill-deserved end many generations ago. We have merely created by means of technology and science a killing capacity that they were obliged to attain by means of cunning and "strategy." The same psychic and moral constellation for creating a society oriented more toward death than life has been with humanity for centuries. Few indeed were the men during this long period who were frightened by their own technical imagination for producing the means for mass slaughter—all legends about Leonardo da Vinci's scruples on this score notwithstanding.

So deep-seated a capacity to use technology's malignant power to destroy instead of its benign power to create requires a searching analysis of the moral elements and origins of what we today call "civilization." Here it suffices to emphasize that the "means of production" have now become too powerful—too manipulable by small, idiosyncratic if not crazed elites, too prolific and cancerous in their metastatic growth—to be designed, much less used, as means of destruction. The steel that the Alexanders and Caesars used to dispatch human life, like the black powder that the Napoleons employed for their artillery bombardments, are puny relics of a relatively benign past. They have been replaced by thermonuclear and neutron bombs, nerve gases, lethal microbes and toxins, and unerring delivery systems that can be used intercontinentally to inflict horrendous destruction by only a few psychologically conditioned human robots—and soon, inhuman robots which can be programmed to "declare" war or peace by men whose own sanity and mental stability are highly dubious.

We are passing the point where technical and scientific advances, apart from mere marginalia, hold any promise for human survival and well-being. With the discovery of nuclear bombs, every technical advance seems to be guided or perverted by the pursuit of increased killing power, its purposes barely concealed by token claims that it is meant to "serve" humanity. Hence it is not mere Luddism to say that we would be safer as a species if we could restore a Paleolithic world of flints than if we were to "advance" to a "post-industrial" world of "intelligent robots." Not that the former is a desideratum in itself, but merely that it is less menacing and demonic in a society ruled by moral cretins and emotional brutes.

III

How did we arrive at a condition where thermonuclear immolation or ecological degradation confronts society as a realistic destiny if we do not recover earlier opportunities to divert the trends of technology, science, and a domineering rationalism so facilely expressed by the word "progress"? Have we merely been mistaken in our judgment of humanity as evolving moral and rational animals moving ever-forward toward the high liberatory ideals of the Renaissance and Enlightenment? Is our species inherently tainted by an irrepressible desire to dominate, to visualize the "Other"—be it nature, woman, ethnic groups, or, quite broadly, our fellow beings—as objects to be manipulated or rivals to be subdued? Is "progress" itself a myth that, by its own self-development, turns into its opposite as regression? Is thermonuclear immolation or ecological degradation the logical fate of a species that has been defective from the start, a species for whom the moral and intellectual trappings of "progress" have concealed a fiercely destructive impulse that has merely found in social evolution the all-powerful tools and means of destruction to tear down the planet?

Or do we have reason to believe that these questions have only limited validity, that they express a sinister departure, comparatively recent in time, from the "mainstream" of human and natural evolution? Do we have reason to hope that human beings have inherently moral and rational qualities that are indeed liberatory, and exist today as a potentiality that can be recovered and realized? Is it a given that the "Other" must be reduced to a mere object of manipulation, or can it exist as an end in itself to be cherished disinterestedly or treated benignly in a caring ecological constellation of living beings? Is there, perhaps, a "mainstream" of progress from which we have diverged like a limb from a tree—an overall movement in the affairs of life and humanity that, lurking in the mists of our history, still holds the promise of new ideals of freedom, love, and highly ethical interaction between human beings and between humanity and nature?

How will we be able to answer these questions, much less understand them, if there are no autonomous thinking people around to formulate them? If we are to answer these questions, we are obliged to step back in history to see if some trend in humanity's evolution held the promise of a truly emancipatory progress. We must try to ascertain if there was another juncture, a branching-off point, of our species and our society from a richly evolving trend toward consciousness and true enlightenment. The dead hand of the past does not lie on the "brain of the living" like a "nightmare," as Marx claimed more than a century ago, nor can sweeping social change "draw its poetry ... only from the future"—for there is no future for a humanity that seems to have veered onto a path that threatens its very survival in the absence of radical change in its institutions and sensibility. Quite to the contrary: the "tradition of all the dead generations" which Marx, in his effluvium of nineteenth-century progressivism, hoped to exorcise with the "poetry" of "the future" has yet to be recovered and explored in the light of the dead-end that confronts us. The future as we know it today, whether in the form of socialism or capitalism, has no poetry to inspire us.[5]

The "dead" have not "buried the dead," as Marx had hoped; their remains surround us for good or evil and provide the examples—both good and evil—by which to judge the present and literally recreate a future based on continuity with a humanistic and ecological past, one that will yield a humanistic and ecological society in the century that lies ahead.[6] Hence we must go backward in time, at least in our consciousness, to determine when and how we "erred" so that we may then regain a lost path that can lead us to a liberatory society.

Perhaps the most comprehensive summary we have of humanity's capacity for aggression and domination, Erich Fromm's *The Anatomy of Human Destructiveness*, effectively refutes images of our species as inherently oriented toward rivalry and subjugation. After an exhaustive review of ethological, paleontological, anthropological, and early historical data, Fromm concludes that human "destructiveness," indeed rivalry, "is neither innate, nor part of 'human nature,' and it is not common to all men." Quite to the contrary: based on an analysis of some thirty Indigenous tribes, the majority are at best "life-affirmative societies" or "non-destructive aggressive societies," which "are by no means permeated by destructiveness or cruelty or by exaggerated suspiciousness," although at worst they may lack "the kind of gentleness and trust" Fromm finds in "life-affirmative societies."[7]

Politically, in a sweeping and perceptive judgment of humanity's extended periods of social development, Jane Mansbridge in her remarkable and well-documented study of democracy, *Beyond Adversary Democracy*, emphasizes that face-to-face, egalitarian, and consensual democracy ("unitary democracy") based on friendship, in contrast to modern "adversary democracy" structured around representation, hierarchy, and majority rule (itself based on competing interests), "almost certainly has a longer history than any form of government. For more than 99 percent of our history, we human beings lived in hunter-gatherer bands, which in all probability practiced unitary democracy."[8]

If these conclusions are sound—and they can now be supported by a considerable amount of data—we must ask ourselves when and how "civilization" veered away from a largely pacific, egalitarian, and caring ordering of human relationships toward an increasingly aggressive, hierarchical, and adversarial social order—that is, toward societies that have tainted almost every major human achievement with destructive powers and demonically elicited from them their potentiality for coercion and domination. Where were the junctures at which human beings began to employ even seemingly benign technologies, reasoning power, and institutions for oppressive and exploitative ends?

Two periods of social evolution that suggest where the branching-off took place seem to emerge from the mists of the past—one, decisive for the course of human history in general, the other, for the formation of the modern era in its most savage and biocidal forms. The primary juncture involves a sequence of shifts from matricentric to patricentric societies, more specifically, from egalitarian and domestically oriented relationships to hierarchically and politically oriented relationships. These reached their crucial moment with the emergence of the bronze-age warrior. The other, more recent branching-off occurred with the discovery of the "New World." The expanding market economy that developed within late feudal society—rarely benign but, as yet, far from devastating in its social and ecological impact—acquired a rapacity, cruelty, and degree of destructiveness unparalleled by any commercial society before it. Beckoned by the precious metals and wealth of two vast continents—the Indian Americas—European society was thrown into a frenzy of greed and insatiable lust for riches that constituted a complete revolution in human values, goals, and needs—a revolution that still permeates our contemporary culture. Even more than Circe in the *Odyssey*, who turned men into beasts, the virgin Americas became a trough that turned Europeans of all nations into swine. We have not shed the bristles of these legendary beings. Worse, we have developed even sharper teeth and more insatiable appetites than the austere Puritans and tempestuous conquistadors who began to cross the Atlantic five centuries ago.

The factors that drove humanity from an egalitarian world into a hierarchical one, from woman's domestic hearth to man's military battleground, from a sensibility rooted in mutualism to one rooted in rivalry are too complex to examine in detail.[9] Moreover, surrounded as they are by thick archaic mists, these changes apparently affected only a small fraction of humanity, particularly cultures in the Near East and portions of the Americas. The rest of humanity, sifted out by a process of negative selection, had to be pushed into what we call "civilization" kicking and screaming with revulsion. It suffices to say that once the patriarch dissolved out the mother-imagery of early society with his growing powers of life and death over the clan, and once the warrior acquired control over the material means of life as a feudal or kingly land-magnate—at first as defender of the community from hostile aliens, later as aggressor and expropriator of the community's lands—the world of hierarchy and domination began to permeate the world of an egalitarian and ecological society. Initially, by eroding earlier sensibilities of complementarity and mutual aid, an earnest respect for nature, and the use of the gift as a token of human solidarity; afterwards, by inverting cooperation into competition in the very course of manipulating the traditional form of communal labor to serve the "megamachine" of mass corvée labor, and by degrading the ancient view of nature as subject into a world of "objects" or "natural resources"; finally, by substituting the exchangeable commodity for the solemn gift—through all these inversions and changes, the phoenix of rivalry, already latent in the "Big Man" syndrome of tribal society, rose from the ashes of complementarity and reciprocity. Humanity's richly textured ties of mutualism were turned into the chains of the insensate buyer-seller relationship in which the bargain, with its trappings of profitability, replaced the simple giving and taking of things according to need.

In a world that is fairly innocent of greed and hierarchy—a world in which the very word "freedom" is absent from the vocabulary because it is a universal reality of life—only a far-reaching consciousness of the ills that emerge with the first breaches of its libertarian "social compact" can prevent the logic of domination from totally altering a community's fragile sensibility of mutual aid and respect for human beings and the natural world. Naivete bears not only the charm of purity, but also a dangerous vulnerability to manipulation. Our children pay this harsh penalty daily as they are "socialized"—and no less was it paid by early human society for its sequence of "elders," patriarchs, warriors, priests, and finally, chieftains, kings, and emperors.

The very troughs that turned men into swine, however, contain the nutrients for armoring men against swinishness. The long and bloody toil of "civilization" over the past ten millennia has brought humanity to the brink not only of mass self-destruction, but quixotically, in its own tortured way, to the brink of acute self-consciousness. The epochal surges of history have removed us from the parochial tribalism of the kin-group into a shared sense of universal humanity in which the torment of peoples far removed from our own community can evoke empathetic sentiments and militant action. The step forward from a self-enclosed folk to a worldwide sense of *humanitas* has added a new species-dimension of concerns to our personal and local concerns.

So, too, lying amidst the technologies of destruction are the technologies of creation that can recover terrains we have already damaged, technologies which can possibly be placed in the service of humanity and nature to foster, rather than degrade, social and natural evolution. The unparalleled opportunity of choosing our own needs, rather than simply bending under the material burden of arduous toil and the lack of means for survival, opens a historically new horizon of time to live creatively, to fulfill our personal and social potentialities as human beings, to lift the sense of "scarcity"— be it mythic or not—from our minds, and to become dignified, secure, and self-assured beings.

Finally, behind us lies the wealth of history itself, the treasure-trove of knowledge—of successes laden with promise and failures laden with fault. We are the heirs of a history that can teach us what we must avoid if we are to escape immolation and what we must pursue if we are to realize freedom and self-fulfillment. We can discover only too easily what our distant ancestors could never know—the trickery and cunning of elites and power-brokers who induced them to chain themselves in servitude. We can undo not only the chains that bind our limbs, but also the chains that bind our minds. New choral notes answer the trumpets of war and the drumbeats of mass cultures: the notes of a totally emancipated world, free of sexual and gender oppression, of ethnic prejudice and ageist neglect, of competitive relationships and the unceasing war between human beings and between humanity and nature, of disempowerment in social life and personal life, of a crudely simplified and brutish selfhood and sensibility—the diminution of people to "human resources" that forms the counterpart of nature conceived as "natural resources."

Such a heritage—indeed, the armor to rescue ourselves from leaders who betray, institutions which enslave, methods which coerce, and sensibilities which domineer—involves the searching study of where we "went wrong" in the course of social evolution ages ago and in more recent times. What we can ascertain from this study is that our gravest departures from the development of humanity and nature toward wholeness and fulfillment are not institutional alone. Coercive as were the temples that replaced the natural shrines and groves of tribal peoples, the brutalizing factories and the lifeless technological imagination that replaced more humanly scaled machines and an aesthetic vision of production, the institutional legacies of domination, whether family or State, that replaced the institutional legacies of freedom—the fact remains that the greatest and most effective modes of coercion stemmed from the prejudices, sense of self, and modes of rationality that brought us into complicity with our self-degradation and the degradation of nature. The internalization of hierarchy and domination forms the greatest wound in human development and the most deadly engine for steering us toward human immolation.

Our step back into the past—all the better to clearsightedly view our present and future—yields the damning conclusion that we are the unknowing architects of our own servitude. What this means, above all, is that the "revolution" which must "draw its poetry ... from the future" must indeed more thoroughly revolutionize humanity than it could have projected a century or even a generation ago. Domination and hierarchy, internalized as sexism, ageism, a manipulative rationality, an envious hatred of other human beings, a passion to "master" nature, and a grasping egotism that arrogantly passes for "individualism" must be exorcised together with the externalized forms of domination and hierarchy that bring "masses" to their knees—or to their mass graves. It is not that past "revolutions" took their inspiration from the depths of the past, but rather that they never went far enough in searching out the depths of the present—the armored sensibilities of rule that yield the fully sculpted institutions of rule.

Domination, be it of nature or human beings, thus unites the great themes of our times: feminism, ecologism, alternative technologies, peace, material security, self-empowerment, community, holistic health, mutual aid, and a sensibility of respect for human beings of all ages and ethnic backgrounds. All are united into a common and coherent focus which we may best call social ecology in its broader theoretical aspects and a libertarian municipalism in its function as a new social practice. Reinforced intellectually and spiritually by this cohering focus, we can trace back the tormented history of our species from its movement away from the preindustrial world opened by the discovery of the "New World" and still further back from its break with a nonhierarchical world that opened with gerontocracies, patriarchy, and the armored warrior of the Homeric world.

In so tracing back these tangential developments which now threaten our very survival, we can learn from the gross distortions they produced in our own sensibilities and social institutions how to formulate the means for retrieval and advance of our consciousness and practice for achieving a free, ecological society.

IV

If we are obliged to recover our humanity in order to rescue it physically, we must ask unequivocally what methods, forms of organization, and institutions we must develop that will bring us from "here to there," from a society that faces biocide to a society that will fulfill humanity's full potentialities.

The gravest single illness of our time is disempowerment. Even the most media-saturated spectator senses, however dimly or intuitively, that she or he has no power over the events and forces that determine humanity's future. A crude jingoistic "patriotism" may merely signify acquiescence to the superhuman powers that be, a self-deceptive desire to follow in the tow of visible institutional and military strength. It does not exhibit any real conviction that, as mere follower, one has acquired even a scintilla of command or shares in its glories. The ruling elites of the world have killed as many of their devotees as their opponents in wars of conquest or repression. The military graveyards provide no less mute testimony to the tribute paid by the obedient than do the debris of concentration camps to the penalty paid by opponents and "social undesirables." Hence the most jingoistic of "patriots" can hardly claim to have a "fatherland" in his or her own soul. Only the myth that "power concentrated in the hands of the few is shared by the many" constitutes the balm for an ever-suppurating ulcer of subjugation that the obedient must share, not unlike the soldier whose uniform is both the badge of might and its target on the battlefield.

But to recognize that reempowerment of the individual is one of the most crucial issues of our era raises the question of the ideologies that profess to confer it and the institutions that are meant to achieve it. Here, the so-called revolutionary ideologies of our era—socialism and even canonical anarchism—fall upon hard times. They can be as deceptive in forming a new consciousness as the conventional ideologies of ruling elites. Socialism and canonical anarchism—the "isms" of *homo economicus*, of "economic man"—were born with the emergence of commercial and industrial capitalism. And however oppositional they may be, their underlying assumption that the wage worker is inherently subversive of capital tends in varying degrees to form the counterpart of the very system they profess to oppose. What is perhaps even more ironic today is that their "constituency" is literally being "phased out" with the very industrial structures and industrial classes whose historic aspirations they hoped to voice. The factory itself, not to speak of its industrial proletariat, is being placed on the block of cybernetics and robotics, just as the yeoman-farmer was placed on the block of industrial agriculture and agribusiness two generations ago. If workers' movements of all kinds are today becoming mute, it may be that the vocal chords of the society which cradled them are disintegrating and they can say nothing new in a world whose very vocabulary of change is altering profoundly.

In fact, the socialist and syndicalist proclivity for economic reductionism is now actually obscurantist. It not only shares in the bourgeois tendency to render material egotism and class interest the centerpieces of history, it also denigrates all attempts to transcend this image of humanity as a mere economic being—indeed, as "man the toolmaker"—by depicting them as mere "marginalia" at best, as "well-intentioned middle-class ideology" at worst, or sneeringly, as "diversionary," "utopian," and "unrealistic." "Feed the face, then give the moral," Bertolt Brecht's coarse and contemptuous image of humanity, bears the stigma of the very corruption of radical ideologies that has marked bourgeois society's capacity to infiltrate every area of social life with its contagion of egotism.

Capitalism, to be sure, did not create the "economy" or "class interest," but it subverted all human traits—be they speculative thought, love, community, friendship, art, or self-governance—with the authority of economic calculation and the rule of quantity. Its "bottom line" is the balance sheet's sum, and its basic vocabulary consists of simple numbers. Insofar as men and women wrestle with the system within its economic parameters, they can no more search beyond the issues raised by "class interest" than the entrepreneur can search beyond the issues raised by profit and capital accumulation. Worker and capitalist remain wedded to each other in a community of shared sensibilities and roles, just as hostile mates are bound to each other by a mindless reverence for "the family" and the anxious demands of "the children."

Ironic as it may seem to radicals who have been pastured on the economism of class war, technological growth, and "scientific socialism," be it libertarian or authoritarian, politics must now acquire a "supremacy" over economics, ethics over material interest, the claims of life over the claims of survival if any effective movement for radical renewal and change is to be achieved. Not that they must be counterposed to each other, but that each must be given its proper weight in the new balance of social events. The so-called superstructure, to use the language of Marxian "historical materialism," cannot be seen as an epiphenomenon of the "base" if there is to be any resistance to a market-oriented society that tends to make the ordinary individual, even the self-anointed revolutionary, into a mirror-image of itself. To the extent that *homo politicus* can replace *homo economicus,* and *homo collectivicus* can replace both, humanity still has a chance to rescue itself from a spiritual catastrophe in which social immolation and ecological breakdown will merely provide the shroud for an already decaying corpse. What this reconnaissance into the real opening of American life means is that all thinking people must participate consciously in the tension between the American Dream conceived as utopia and the American Dream conceived as a huge shopping mall. This tension is compellingly real. Perhaps no people today is more conflicted over its commitment to individual rights, its freedom from governmental control, and its freedom of expression at one end of the spectrum, and its desire to surrender all public responsibility and autonomy to the siren-call of material comfort and the opiates of electronic mass culture at the other end. Within this area of conflict the Right has enjoyed a monopoly of power unmatched by a largely indifferent Left. Herein lies a tragedy of monumental proportions. Still nourished by a classical image of radicalism and tormented by a searing guilt for America's role in the so-called "Third World," American radicals have remained strangers in their own homeland. Had the Congregationalist town-meeting conception of democracy been fostered over aristocratic proclivities for hierarchy, had political liberty been given emphasis over *laissez-faire, laissez-aller*; had individualism been given an ethical ideal instead of congealing into proprietary egotism; had the Republic been slowly reworked into a confederal democracy; had capital concentration been inhibited by cooperatives and small, possibly worker-controlled enterprises; had the rights of localities been rescued from the centralized state; and finally, had the middle classes been joined to the working classes in a genuine people's movement such as the Populists tried to achieve (instead of being fractured into sharply delineated class movements)—in short, had this American vision of utopia supplanted the Euro-socialist vision of a nationalized, planned, and centralized economy, it would be difficult to predict the innovative direction American social life might have followed.[10]

This does not mean this tradition cannot be reclaimed. Critically important today is the need to recreate a democratic public, a body politic committed to the ideals of free expression and the right of every person to formulate social policy. Contemporary urban society and the mass media have subverted the very notion of an active citizenry—the soul of such a body politic. Its decline can be summed up in the most fundamental feature of American life: disempowerment.

Reempowerment presupposes that every individual can feel he or she has control over the decisions that affect our society's destiny. Such a sensibility can only be recreated and fostered by a radical change in the scale of everyday life, a conscious endeavor to bring the social environment within the purview of the individual, to render it as comprehensible and understandable as possible. No mere intuitive actions and explosive episodes will do for a society that threatens to replace "primitive" innocence and naïvete with a "sophisticated" cynicism and indifference. Decentralization of decision making, and the institutionalization of the "grass roots" into impregnable structures that are built on a face-to-face democracy, constitute the unavoidable challenges that can form a new, active citizenry in a real participatory democracy. Without this "unitary democracy," democracy of any kind may well disappear. Corporate America cannot assert itself over the existing and potential means of power by hybridizing even republican institutions with totalitarian ones. We do not know whether anything less than complete state control can avoid any threat to the corporate order as such and the deployment of a stupendous technological armory to deal with the problems of a historically new epoch in human relationships.

In the United States, despite the erosion of our libertarian institutions over the past few generations, American political life still bears the deep imprints of its revolutionary origins. Mythic or real, individual rights, juridical equality, free expression, and resistance to State encroachment still exercise a powerful hold upon the American mind. This is a genuine reality in its own right, all losses of such values notwithstanding. The ideal of "liberty," however varied its meanings to different citizens, looms over the Caesarist challenge to the Republic as a haunting memory that has yet to be uprooted from our national heritage and our political conscience. *Homo economicus* has not yet completely supplanted *homo politicus*. The "Bill of

Rights" and the demands of the great Declaration of Independence for "life, liberty, and the pursuit of happiness" still beleaguer the authoritarians like a ghostly army from the past—a *living* past that can yet be galvanized to recreate an empowered, active citizenry and a democratic body politic.

But how can these ideals be given a palpable form at the base of a society already highly centralized politically and economically? Indeed, a society riddled by spectatorial "citizens" who seem like fair game for the mass media and the political star system? We encounter here the problem of recovering or revitalizing forms of democratic practice already in existence which lie dormant in a political community. I refer to forms that are still structured round the idea of decentralization and human scale—notably, the municipality as the ultimate source of power, be it the neighborhood assembly in large cities or the town meeting in small communities. The United States has given greater moral authority than any other country today to the "grass roots," a distinctively American expression that stems from our traditional emphasis on local government and our uniquely libertarian revolution. If a radical practice of public reempowerment is to take itself seriously, it must initiate the act of reempowering the citizen in the environment in which he or she is most directly immersed—the neighborhood or town. On this basic level of political and social life, it must try to create at least exemplary forms of public assembly whose moral authority slowly can be turned into political authority at the base of society. It may not be given that such a sequence of steps is practical in every American municipality, much less every region of America. But where it is practical or even remotely possible, it must become the most important endeavor of a new radical populism—a new libertarian populism.

The term "libertarian" itself, to be sure, raises a problem, notably, the specious identification of an anti-authoritarian ideology with a straggling movement for "pure capitalism" and "free trade." This movement never created the word; it appropriated it from the anarchist movement of the last century. And it should be recovered by those anti-authoritarians—whether socialistic or anarchic—who try to speak for dominated people as a whole, not for personal egotists who identify freedom with entrepreneurship and profit. It would be wiser to simply ignore this specious movement for "liberty" and restore in practice a tradition that has been denatured by new disciples of Adam Smith.

Lest the word "libertarian" be seen as a political capitulation to statism, we would do well to realize a flaw in the authentic libertarian tradition that confuses politics in its Hellenic sense with statecraft. The traditional libertarian counterposition of "society" to "the State" is not false as such. Social forms like families, clans, tribes, guilds, workshops, village communities, neighborhoods, and towns are the organic institutional forms by which humanity "naturally" developed toward consociation and by which it metabolized with nature in the form of production. It is within these forms that the great anarchic theorists hoped to structure a confederal libertarian society. The State, quite soundly, was seen as an exogenous institution—a professionalized class instrument of executives, legislators, bureaucrats, soldiers, judges, and police with their paraphernalia of barracks, courts, and prisons. This exogenous institution had to be consistently bypassed in daily social activity and disbanded in periods of sweeping social change.

For the present, however, what should be emphasized is that the historical landscape is composed of more than society and the State. We must move beyond this simplistic and Manichean dualism to focus on a generally unexplored arena of human activity, a public space or *political* arena in the classical Hellenic sense of the word *politika,* or activity of the *polis,* that cannot be subsumed by the word "State," much less "city-state." Perhaps for the first time in history, but by no means the last, the Athenian democratic *polis* produced an entirely new institutional arena—an arena that was not specifically "social" like the family, clan, tribe, presumably workplace, and various "natural" forms of consociation like clubs, cultic communities, vocational collegia (later, "guilds"), and professional societies. Nor was this institutional arena identical to—or even an extension of— that class-controlled professionalized system of violence such as armies, police, bureaucrats, judges, legislators, and centralized executives we call "the State"—a constellation of structures which were few enough in Periclean Athens although they were quite common in the ancient world.

What the Athenian *polis* created was a uniquely *civic* sphere—a distinctly municipal arena—characterized by the *agora*, or civic center, where citizens could gather informally, discuss, trade, and engage in a richly textured interaction that prepared them for the weekly meetings of all the citizens in the *ecclesia*, or popular assembly, where they normally discussed the issues of the *polis* with a view toward arriving at a public consensus in a face-to-face manner, either with unanimity or by a vote. There, too, as in the *agora*, they were daily and subtly educated into the arts and attributes of active citizenship, or more precisely, into the sensibility, character-structure, and selfhood of participatory and self-governing citizens—a new, specifically municipal "class" in the spectrum of "classes" which Marx and our current bouquet of radical social theorists have so neatly arranged for us. The Athenian democracy, in effect, functioned as a school for personal and social development, and with its dramas, festivals, and pageants provided a unifying *cultural* milieu that knitted together the *polis* into a community unified in sensibility and tradition. Economic strata Athens had in abundance: slaves; craftpersons; merchants; alien residents, or *metics*, who enjoyed none of the political rights of citizens—although there were free; independent farmers; tenant farmers; intellectuals who practiced what we today would call "professions"; nobles of aristocratic lineages; and, finally, a stratum of demagogues who tried to manipulate the citizen body to suit their own personal and social interests.

But there was also the citizen body itself—the body politic—that, despite its mixed "class" character, often transcended its particularistic economic interests to arrive at an ethical consensus. This consensus was guided by a notion of the "public good," a "good" that cannot be dismissed as purely "ideological" or subtly class motivated, but which rested on a shared notion of what was transparently the public welfare, all particularistic frictions and conflicts aside. As expressed by the social philosophers of Greece, whether democratic or authoritarian, this notion of the "public good" consisted of the body politic's *arete*, or virtue. In the Athenian democratic *polis*, politics, or *politika*, to use a safely Hellenized term, belonged neither to the realm of the "social" nor that of the "State," terms for which the Greeks significantly had no words. It comprised the realm of the *polis*, of the *civis*, to use a more familiar Latin word, in which men (the society, like all those around it, was firmly patricentric) formed a "social compact" or common ethical understanding—in no way to be confused with a juridical "social contract"—to order its life as a "commonweal" to try to transcend particular interests.

What the anarchic theorists have not seen clearly is a supra-social level of politics—literally, the activity of the polis—that can validly be distinguished from *statecraft*. Politics is the public realm of citizenship where citizens gather to discuss social problems, evaluate them, and, finally, decide on their solution, whether by consensus or by vote. This political arena, as distinguished from the largely social world of organic relationships at one end of the spectrum and the statist world of ruling-class controls at the other end, is the intermediate world of the community and the citizen—a municipal world based not on kin but on the *civic* association that so often surfaces in the writing of Proudhon and Kropotkin, only to be overlaid by the industrial world of syndicalism.

Tragically the liberatory side of community and municipal politics was never fully elaborated. Indeed, it was myopically subsumed by an "antipolitical" bias that brought the distinctions caused by statist and social infiltration into a twilight zone of radical social theory. Libertarian municipalism was further stigmatized by the crude betrayals of its most outstanding spokesman, Paul Brousse, the French anarchist who drifted in his later life toward conventional party politics. Cleansed of this stigma, libertarian municipalism *and* its politics, based on popular assemblies and confederal relationships, can be seen as a process that does not deny politics in its *classical* sense but rather serves to give it authenticity and contemporary relevance.

This perspective—a libertarian populism based on municipal freedom and confederation—is different from those which normally appear in most radical social theories. It speaks more directly and traditionally to American conditions than the European traditions of radicalism. It makes no attempt to imperialize the ideological landscape—and it carries the warning that ages long past, both in American history and European, may be irrecoverable for people whose very spirit may be industrialized and reduced to spectatorial passivity.

If nothing else, however, it tries to speak to a more independent, politically concerned, and libertarian American spirit that may still lie latent in the national character-structure. And if nothing is left of that spirit, radicals may follow their own personal course, perhaps returning to the daydreams of dignity in defeat with the certain knowledge that defeat will end in biocide. But if something of that spirit is still left, even as embers, these remarks may be regarded as part of an effort to raise up the flames of protest and provide us with the means for reconstruction.

Here, at least, is a chance for humanity to regain its sanity and rebuild this ruined planet as a world for life. It is a chance that must arouse the very unconscious of the individual and redeem the spirit of life with which he or she was born to produce a new culture and consciousness, not only a new movement and program.

March, 1983

Notes

Rethinking Ethics, Nature, and Society

1. G. P. Maximoff, ed., *The Political Philosophy of Bakunin: Scientific Anarchism* (Glencoe, NY: The Free Press, 1953), 358.
2. Let me emphasize the word *ground* and point out that I do not use the word *source*, which has been carelessly tagged onto my views. Ethics presupposes the presence of volition, the intellectual ability to conceptualize and the social ability to institutionalize communities, not merely to collect into a community. These capacities are uniquely human and deserve emphasis as such, all the more because certain environmentalists tend to ignore humanity's uniqueness as a potentially rational and social animal in the sense that I use the word "social." Nature does not have will in the human sense of the term, nor does it possess the power of conceptualization. The sense in which nature is a ground for ethics, not ethical as such, will be explored shortly.
3. G. W. F. Hegel, *The Phenomonology of Mind*, trans. J. B. Baillie (New York: Humanities Press, 1910), 79.
4. This is as good a place as any to point out that process-thinking of this kind is meaningless without empirical verification of our conclusions in the real world. But an empirical test of our ideas cannot be an attempt to test a fixed set of ideas with a fixed set of facts, as the acolytes of common sense and a crude pragmatism would have us believe. In truly dialectical thinking, an empirical test must explore whether a given process in its theoretical form explains a given process in real life. That the processes in thought must try to explain

processes in reality is a basic notion of empirical verification that has eluded even many self-professed "dialecticians" and left them open to compelling criticisms by pragmatists and positivists.

5. These analogies and the thoughts behind them have been so freely lifted from my writings, frequently with minimal or no acknowledgment, that the reader should be mindful of where they initially originated and the contexts in which they originally appeared. My concern is not simply with the problem of plagiarism; rather, I am much more troubled by the way they are used to support ideologies with which I differ profoundly, such as notions of "natural law" that provide a warrant to use authoritarian methods for the correction of ecological dislocations. In this view of the libertarian message of social ecology, I find this hybridization of ideas very disturbing. I feel that a cautionary note should be made in the interests of freedom and clarity of ideas.

6. Karl Marx, *The Grundrisse*, trans. David McLellan (New York: Harper & Row, 1972), 94.

7. See particularly *Post-Scarcity Anarchism*, Ramparts Press: which has been reprinted by AK Press, and *The Ecology of Freedom* (Palo Alto: Cheshire Books, 1983, reprinted by AK Press). Readers may also care to consult *Toward An Ecological Society*, also published by AK Press.

8. Jules Michelet, *History of the French Revolution* (Chicago: University of Chicago Press, 1973), 444.

9. And I would emphasize that more than ever, today, we need a new movement for moral reawakening, not only for meeting human material needs—important as these are at all times. The great failing of contemporary "Leftist" movements, be they socialist or anarchist, is that a new society is conceived primarily as one that places supper on the table, with the ironic result that the conservative Right has gained the support of millions through moral appeals that give a sense of meaning to life in an increasingly meaningless society. I am thoroughly convinced that no new social movement will capture the imagination of people today without providing a sense of moral well-being, not only material well-being—indeed, of moral purpose, not only material improvement.

10. The transclass nature of municipal movements is clearly revealed in Manuel Castell's brief analysis of the occupational background of arrested or deported Communards after the Paris Commune of 1871. It may be well to note that the Commune was universally regarded as a "working class" insurrection and "model" for a Marxist "proletarian dictatorship." Historical material on the opposition which the nation-state encountered from cities can be found in Chapter Six of my book, *From Urbanization to Cities* (Chico, CA: AK Press, 2021). See

Manuel Castell, *The City and the Grassroots* (Berkeley: University of California Press, 1983), 16–17.

11. Jean-Jacques Rousseau, *The Social Contract* (New York: E. P. Dutton & Co., 1950), 94.

12. Editor's Note: The campism by some on the Left that has led to justification more recently of Russia's support for the Assad regime in Syria, which included the deliberate bombing of hospitals, schools, and civilians and a similar reluctance to criticize Putin's invasion of Ukraine, also falls into this category of knee-jerk uncritical "solidarity" with anti-American regimes even when they are themselves a model of imperialism, empire-building, and flagrant opposition to movements for democracy.

What is Social Ecology?

1. I am not saying that complexity necessarily yields subjectivity, merely that it is difficult to conceive of subjectivity without complexity, specifically the nervous system. Human beings, as active agents in changing their environments to suit their needs, could not have achieved their present level of control over their environments without their extraordinary complex brains and nervous systems—a remarkable example of the specialization of an organ system that had highly general functions.

2. Neil Evernden, *The Natural Alien* (Toronto: University of Toronto Press, 1986), 109.

3. Quoted in Alan Wolfe, "Up from Humanism," *American Prospect* (Winter 1991): 125.

4. Paul Radin, *The World of Primitive Man* (New York: Grove Press, 1960), 211.

5. Murray Bookchin, *The Ecology of Freedom* (Oakland: AK Press, 2005), 94.

6. In its call for a collective effort to change society, social ecology has never eschewed the need for a radically new spirituality or mentality. As early as 1965, the first public statement to advance the ideas of social ecology concluded with the injunction: "The cast of mind that today organizes differences among human and other life-forms along hierarchical lines of 'supremacy or 'inferiority' will give way to an outlook that deals with diversity in an ecological manner—that is, according to an ethics of complementarity." Murray Bookchin, "Ecology and Revolutionary Thought," originally published in the libertarian socialist periodical *Comment* (September 1965) and collected, together with all my major essays of the 1960s, in *Post-Scarcity*

Anarchism (Berkeley: Ramparts Press, 1972; reprinted Oakland: AK Press, 2004). The expression "ethics of complementarity" is from my *The Ecology of Freedom: The Emergence and Dissolution of Hierarchy.*

Indeed, such a change would involve a far-reaching transformation of our prevailing mentality of domination into one of complementarity, one that sees our role in the natural world as creative, supportive, and deeply appreciative of the well-being of nonhuman life. In social ecology, a truly *natural* spirituality, free of mystical regressions, would center on the ability of an emancipated humanity to function as ethical agents for diminishing needless suffering, engaging in ecological restoration, and fostering an aesthetic appreciation of natural evolution in all its fecundity and diversity. In such an ethics, human beings would complement nonhuman beings with their own capacities to produce a richer, creative, and developmental whole—not as a "dominant" species but as a supportive one. Although this ethics, expressed at times as an appeal for the "respiritization of the natural world," recurs throughout the literature of social ecology, it should not be mistaken for a theology that raises a deity above the natural world or even that seeks to discover one within it. The spirituality advanced by social ecology is definitively *naturalist* (as one would expect, given its relation to ecology itself, which stems from the biological sciences) rather than supernaturalistic or pantheistic areas of speculation. At a time when a blind social mechanism—the market—is turning soil into sand, covering fertile land with concrete, poisoning air and water, and producing sweeping climatic and atmospheric changes, we cannot ignore the impact that an aggressive hierarchical and exploitative class society has on the natural world. We must face the fact that economic growth, gender oppressions, and ethnic domination—not to speak of corporate, state, and bureaucratic incursions on human well-being—are much more capable of shaping the future of the natural world than are privatistic forms of spiritual self-redemption. These forms of domination must be confronted by collective action and by major social movements that challenge the social sources of the ecological crisis, not simply by personalistic forms of consumption and investment that often go under the oxymoronic rubric of "green capitalism." The present highly cooptative society is only too eager to find new means of commercial aggrandizement and to add ecological verbiage to its advertising and customer relations efforts.

7. *Der Spiegel,* September 16, 1991, 144–45.
8. I spelled out all these views in my 1964–65 essay "Ecology and Revolutionary Thought," and they were assimilated over time by subsequent ecology movements. Many of the technological views advanced

in my 1965 essay "Toward a Liberatory Technology" were also assimilated and renamed "appropriate technology," a rather socially neutral
expression in comparison with my original term "ecotechnology."
Both of these essays can be found in *Post-Scarcity Anarchism*.

9. See "The Forms of Freedom" in *Post Scarcity-Anarchism*; "The Legacy
of Freedom" in *The Ecology of Freedom*; and "Patterns of Civic Freedom," in *From Urbanization to Cities: The Politics of Democratic Municipalism* (Chico, CA: AK Press, 2021).

Market Economy or Moral Economy?

1. Theodor Adorno, *Minima Moralia* (London: New Left Books, 1974), 156.
2. Marx, like David Ricardo, played a major role in divesting economic
theory of its moral content and surrounding it with a scientistic
ambience even while he denounced capitalism for its brutality and
egotism. Marx's *Capital* is riddled with mixed messages that impute
the all-presiding, seemingly "just" role to equivalence in the capitalist economy, particularly in the exchange of labor power for wages,
while exhibiting a genuine revulsion for an economic system that
reduces every human relationship to a cash nexus. Marx's scorn for
demands like "economic justice," particularly a "just wage," seems to
be almost unknown to most Marxists these days, a scorn which would
be laudable were it not the product of his own scientistic image of
economics as the study of "the natural laws of capitalist production,"
Karl Marx, *Capital* (New York: Modern Library, 1906), 13. For further
discussion of the nature of justice, see Chapter V of *The Ecology of
Freedom* (Palo Alto: Cheshire Books; 1982; revised edition with a new
introduction, Oakland: AK Press, 2005.)
3. The notion of usufruct, the freedom of individuals to appropriate
resources merely because they want to use them at a time when the
"owner" has no need of them, is too complex to discuss here. For a
more thorough and historical examination of the principle, see *The
Ecology of Freedom*, especially pp. 116–18.
4. This function has often been sadly overlooked by many food cooperatives which, for a time, were administered by the "cooperators" who
did the buying as well as the "staff" which organized the distribution
of food. That the need for "efficiency" and the competitive stance in
which many such cooperatives were placed with large commercial
food emporia ultimately provided some justification for a "tighten-
ing up" of their operations goes without saying. What is troubling,
however, is that the mentality which the seemingly more concerned
administrators of the cooperatives exhibited often differed very little

from that which we would expect to find in the manager of a super-market. "Efficiency" was not merely placed before morality and the educative functions of a food cooperative; the latter simply dropped out of sight completely, as though a food cooperative was a *cheaper* depot for victuals rather than a *cooperatve* in any sense of the term.

Workers and the Peace Movement

1. A caveat that cannot be repeated often enough: that workers are often militant does not mean that they are revolutionary. The American "Left" still lives on the imagery of the 1930s CIO factory sit-ins and strikes. I can personally attest to the fact that while these were very militant waves of action, they never threatened the existence of American capitalism. The American flag was the most prominently displayed symbol of the great worker demonstrations—and by worker demonstrations, I do not mean those orchestrated by the Communists or socialists of the period. No one would have survived a public oration at such demonstrations that challenged the system as such—merely its abuses and the need to reform it. We must dispel the myth that the United States was on the point of "revolution" at any time during the 1930s or that the union-organizing movements that brought millions of industrial workers into action constituted any danger to the existence of capitalist society. Many individual workers may have generalized from their experiences beyond mere reform-ist lines, but the class as a whole could always be contained within the economic and political status quo. They demanded their "right-ful" place within the system, not its revolutionary change. And their movements were to ebb as their leaders and privileged sectors of the labor movement acquired a "rightful" place in a society based on the supremacy of capital.

2. It is ironic that Lenin in his famous *What Is To Be Done?* assigned this "declassé" status precisely to members of the Bolshevik Party, who were expected to transcend their class origins—proletarian or otherwise—and constitute themselves into a supraclass elite called "professional revolutionaries." In this respect, Lenin was acutely conscious of the fact that "The Proletariat" *qua class* would remain captive to "economism" and never arrive at a revolutionary con-sciousness. The democratization of this idea is alien to Marxism on its own terms. Marxism is, by its very nature, economistic, just like bourgeois ideology. Lenin did not depart from this economistic notion; he merely created a privileged elite that would comprise the "vanguard" of a "hegemonic" class, a general staff of an otherwise

"brutish" army. Why such a "general staff" could come into existence in the first place raises a fascinating issue. Even more fascinating is the question of how it could escape the economistic mentality of the very army it professed to be leading. Indeed, herein lay the germs for the degeneration of the Bolsheviks, all other social considerations aside. It was not better—or more democratic—than the class it tried to lead. Indeed, more often than not, the Russian workers, largely peasants-in-overalls, tried desperately to escape from the "economistic" dogma in which the Bolsheviks had straitjacketed them—witness the Kronstadt Commune of 1921—but to no avail, thanks largely to their "general staff."

An Appeal for Social and Ecological Sanity

1. Editor's note: This essay, completed on March 19, 1983, reflects the period in which it was written with respect to country-specific Cold War events. Rather than try to update it, I have opted to leave the author's original voice and message; for while certain country-specific issues may no longer be in play today, it bears eerie witness to the similarity of more recent geopolitical conflicts in the post-Cold War era and, with the Russian invasion of Ukraine, what some have arguably called the start of a new Cold War in 2022.

2. Editor's Note: As with the section above, I've chosen to leave the observations and data regarding ecological destruction in their original form; they stand as a prescient warning from forty years ago of the massive ecological destruction with which we have now become all too familiar.

3. Editor's Note: More recently the World Wildlife Fund has estimated that each year 50,000 square miles of rainforest are destroyed, an area the size of England, annually, https://wwf.panda.org/discover/our_focus/forests_practice/importance_forests/tropical_rainforest.

4. See, for example, my essay "Ecology and Revolutionary Thought," a work written in 1964, later reprinted in my book *Post-Scarcity Anarchism* (Berkeley: Ramparts Press, 1972; reprinted Oakland: AK Press, 2004), which projected many of these problems into the far-distant future. This essay, which seemed so extravagant two decades prior to this one, because of its hypothetical projections, could now be regarded as an understatement of ecological dislocations that currently confront us.

5. The quotations in this paragraph are drawn from Karl Marx, *The Eighteenth Brumaire of Louis Napoleon*, in Karl Max and Friedrich Engels, *Selected Works*, vol. 1 (Moscow: Progress Publishers, n.d.), 400.

6. It is in large part this project of recovery and reexamination that my book, *The Ecology of Freedom* (San Francisco: Cheshire Books, 1982; revised edition with a new introduction, Oakland: AK Press, 2005), was intended to undertake; in it the reader will find a more expansive exploration.

7. Erich Fromm, *The Anatomy of Human Destructiveness* (New York: Holt, Rinehart and Winston, 1973), 181, 169.

8. Jane Mansbridge, *Beyond Adversary Democracy* (New York: Basic Books, 1980), 10.

9. Here again I must refer the reader to *The Ecology of Freedom*, particularly the opening three chapters, for a more thorough account of the way in which gerontocracies, patriarchy, warriors, priestly corporations, and, finally, the State vastly altered the traditional social landscape of humanity, structured around kinship relationships and tribal institutions. Limits of space make it impossible for me to explore the remarkable ways in which egalitarian institutions and nonhierarchical values were turned against themselves in the very process of humanity's movement out of organic and truly non-domineering social and psychological relationships.

10. Global South peoples now face very similar alternatives. Will the Indigenous people in Central America, for example, be free to establish autonomous communities rooted in their rich, pre-industrial, native cultural heritage? Or will they be colonized, ultimately even exterminated, by American-supported feudal juntas and Russian-supported technocratic "Ladinos" who conjointly seek to exploit their labor and resources?

Index

production, 58, 67–8
progress, 122–3, 125–7
proletarian revolutions, 98–100
proletarians. *See* workers/prole-
    tariat
protests, xi–xii
public realm, 78, 79–80, 82–3
public welfare, 136–7

Q
quality of goods, 56

R
racism, 57, 100
radicalism: as alternative, 77–8;
    and capitalism, 2; and commu-
    nities, 87; corruption of, 131;
    ecological movement, 79–80;
    vs. militant action, 88–9; and
    municipalities, 87–8; and
    nationalism, 30–1; replaced by
    managed capitalism, 78–9; as
    schizophrenic, 3; and working
    people, 107–8
Radin, Paul, 44–5
rationalism, 102
reactionary movements, 33
realism, 3, 4–5
reductionism, 10, 40, 130–1
reempowerment, 133
reform, 104, 144n1
religion, 7, 43
renewable energy, 56
representation, 29
respect, 46
revolts, 50, 86, 98–9, 115–16
revolution: American political
    life, 133–4, 137–8; and com-
    munities, 85–6; and German
    working class 1918-19, 103; and

solidarity, 23–4; success of,
    99–100; U.S. in 1930's, 144n1;
    and workers militancy, 98. *See
    also specific revolutions*
risks vs. benefits, 4–5, 7
rivalry, 124
Rousseau, Jean-Jacques, 29
Russia, 31, 114–17, 141n12, 145n1
Russian Revolution, 99

S
Scheidemann, Philipp Heinrich,
    103
science, 119–21, 122
second nature, 39, 40–2, 50
self-domination, 16
self-interest, 1, 3
shopping malls, 66, 69
simplification, 119
slavery, 56–7, 84
Social Democratic Party (Ger-
    many), 103–4
social ecology: overview, 54–9;
    and biological thinking, 10,
    41; and capitalism, 20–1;
    defining, 11–12, 35, 40–1; and
    domination, 54–5; and ego,
    25; and ethics, 8; failure of,
    58; first/biotic nature, 41; and
    free municipalities, 29; and
    growth, 51–2; and harmony, 15;
    and hierarchy, 50, 55, 141n6;
    movement of, 56; and partic-
    ipation, 17; as philosophy of
    potentiality, 9–10; as political,
    19; and politics, 56, 57; and re-
    ductionism, 10; second/social
    nature, 41–2; as sensibility, 17;
    and species hierarchy, 16–17;
    and spirituality, 48, 142n6

Murray Bookchin

One of the most important radical thinkers of the last century, Murray Bookchin (1921–2006) originated a reconstructive social theory called "social ecology," blending aspects of classical Greek and modern philosophy, anarchism, anthropology, and ecology in an effort to rethink humanity's relationship with nature. His groundbreaking essay, "Ecology and Revolutionary Thought" (1964), was one of the first to assert that capitalism's grow-or-die ethos was on a dangerous collision course with the natural world that would include the devastation of the planet by global warming. A long-time activist, and author of two dozen books on ecology, history, philosophy, and urbanization, Bookchin insisted that a complete transformation in social relations, in which all forms of hierarchy and domination were eliminated, was essential if we are to heal our relationship with nature. His work has influenced numerous movements around the world, including the New Left of the 1960s, the alterglobalization movement, the radical municipalism movement, and the Kurdish democratic confederalism project in Turkey and Northeast Syria.

AK PRESS is small, in terms of staff and resources, but we also manage to be one of the world's most productive anarchist publishing houses. We publish close to twenty books every year, and distribute thousands of other titles published by like-minded independent presses and projects from around the globe. We're entirely worker run and democratically managed. We operate without a corporate structure—no boss, no managers, no bullshit.

The **FRIENDS OF AK PRESS** program is a way you can directly contribute to the continued existence of AK Press, and ensure that we're able to keep publishing books like this one! Friends pay $25 a month directly into our publishing account ($30 for Canada, $35 for international), and receive a copy of every book AK Press publishes for the duration of their membership! Friends also receive a discount on anything they order from our website or buy at a table: 50% on AK titles, and 30% on everything else. We have a Friends of AK ebook program as well: $15 a month gets you an electronic copy of every book we publish for the duration of your membership. *You can even sponsor a very discounted membership for someone in prison.*

Email **friendsofak@akpress.org** for more info, or visit the website: **https://www.akpress.org/friends.html**.

There are always great book projects in the works—so sign up now to become a Friend of AK Press, and let the presses roll!